D0758761

SAVING SEA TURTLES

SAVING SEA TURTLES

Extraordinary Stories from the Battle against Extinction

JAMES R. SPOTILA

The Johns Hopkins University Press • *Baltimore*

Printed in the United States of America on
acid-free paper
9 8 7 6 5 4 3 2 1

The Johns Hopkins University Press
2715 North Charles Street
Baltimore, Maryland 21218-4363
www.press.jhu.edu

Library of Congress Cataloging-in-Publication Data
Spotila, James R., 1944–
Saving sea turtles : extraordinary stories from the battle
against extinction / James R. Spotila.
p. cm.
Includes index.
ISBN-13: 978-0-8018-9907-2 (hardcover : alk. paper)
ISBN-10: 0-8018-9907-9 (hardcover : alk. paper)
1. Sea turtles. I. Title.
QL666.C536S688 2011
639.9'77928—DC22 2010023463

A catalog record for this book is available from the British Library.

The Johns Hopkins University Press uses environmentally friendly book materials, including recycled text paper that is composed of at least 30 percent post-consumer waste, whenever possible. All of our book papers are acid-free, and our jackets and covers are printed on paper with recycled content.

To my wife, Laurie, without whose help
I would not have accomplished anything
in my professional life

and

to Frank Paladino, my student, colleague,
and friend, who has been there with me
and done the heavy lifting for many years

CONTENTS

Preface ix

1 Sea Turtles in the Modern World: *Where Do We Stand Today?* 1

2 Life in the Egg: *Buried Alive under Two Feet of Sand* 15

3 Race to the Sea: *Coatis, Crabs, and Night Herons—Oh My!* 36

4 To the Horizon: *The First Day* 52

5 Lost and Found: *Life as a Juvenile* 57

6 The Deadliest Catch: *The Other Side of Fishing* 78

7 Out of the Fire: *The Gauntlet Continues* 93

8 Return to the Beach: *You Can't Go Home Anymore* 100

9 Nesting: *Taking Back the Night* 117

10 Las Baulas: *The Last Hope for Pacific Leatherbacks* 132

11 Ostional: *The Egg-stained Sands of Costa Rica* 144

12 Global Warming: *Rising Seas, Lost Beaches, and Genders* 160

13 The Great Turtle Race: *A New Approach to Conservation Education* 169

14 Sea Turtles and Satellites: *Tales of Technology* 184

15 2100: *A World with, or without, Sea Turtles?* 198

Index 207

PREFACE

There are only a few books that make a real difference in the world. *The Windward Road: Adventures of a Naturalist on Remote Caribbean Shores* by Archie Carr was certainly one of them. It led to a conservation movement on behalf of sea turtles that continues to this day. It inspired a generation of people to study and conserve sea turtles. I always thought that it would be interesting to write a book about the return to the windward road, about what the world of sea turtles is like today and what adventures await the reader on the Caribbean shore. However, I soon realized that you can never go home again. The world of the sea turtle will never be the same as it was in 1956, when Archie wrote that book. In addition, there is no way anyone will write a book as beautifully conceived and executed as the original.

Nevertheless, sea turtles need a voice in every generation. I have been fortunate to work with Archie and several of his students. I have also been fortunate to work with leaders of the conservation movement in Costa Rica like Mario Boza, Álvaro Ugalde, Doña Karen Olsen, María Teresa Koberg, Randall Arauz, and others. I have been blessed to have some of the finest colleagues and students with whom any professor could ever wish to be associated. All of these people have worked to study and save sea turtles, and together we have learned some new and interesting things about sea turtles and have made some progress in saving them from extinction. I hope that this book can be a small voice for the turtles. It discusses their biology within the context of the struggle to save them. It contains facts and stories that will provide information and hope so that people today will begin the work of conservation anew and keep the dream alive of oceans full of sea turtles and beaches covered with hatchlings.

We all stand on the shoulders of those who have gone before us. There is no way that any one person can replace Archie Carr in all that he did and all that he stood for in the sea turtle world. However, if two or three of us work together we can make some progress, and if more people join in we can make a movement. That movement can turn the tide of slaughter in the oceans of the world and development on the nesting beaches. Then we can accomplish what Archie hoped for more than 50 years ago: bringing back the fleets of turtles that Columbus found when he discovered the New World.

The world is a much more complicated and crowded place today than it was in 1956. The problems are greater and more numerous, and we have less time left to solve them. We no longer have the luxury of eating sea turtles and their eggs, of making jewelry out of their shells and leather out of their skin. Old ways have to pass on if sea turtles are to move forward. There are too many people with too many needs and desires to keep up the pretense that sea turtles can be a source of sustainable harvest. Turtles live too long, they mature too old, and they reproduce too little to support use at the same rate per person that they did when the earth held many fewer people. But we still have time. If we work together, if we come together as a family on the nesting beaches, on the oceans, in the meeting rooms, and in the halls of politics, then we will be successful.

Those who come together to walk a turtle beach, to excavate a sea turtle nest to save some hatchlings, to work to stop fishing practices that kill turtles are part of a family. The very act of doing something for the turtles is an expression of faith in something larger than oneself. The reason that I have hope is that there is a large family of people who all do their part to save sea turtles for one more day. That work may be something small or something large. It may gain headlines or it may go unnoticed. It may involve a once-in-a-lifetime trip to a turtle beach or an everyday action. It may involve saving a pond turtle far from the sea or documenting bycatch on a fishing vessel in the middle of the ocean. I feel that I am a member of a great family of believers, believers in the future of sea turtles. You are my brothers and my sisters. Thank you for all that you are doing.

Thank you as well to all of my students and colleagues who have walked with me for the last 33 years of adventure and inquiry. You have inspired me to work harder to discover new knowledge about sea turtles, to educate people about sea turtles, and to work to save sea turtles. It is an honor to be part of your family.

Finally, thank you to all those who have helped make this book a reality. Mark Gatlin did a great job of editing the text on short notice, and Karen Bjorndal, Gabriela Blanco, Alan Bolten, Mario Boza, Laurie Cotroneo, Shaya Honarvar, Susan Kilham, Kenneth Lohmann, Rod Mast, Andrew McCollum, Anne Meylan, Sally Murphy, Carlos Mario Orrego, Frank Paladino, Pilar Santidrián Tomillo, Jeffrey Seminoff, George Shillinger, Annette Sieg, Paul Sotherland, and Edward Standora reviewed various chapters, and in some cases the entire manuscript, to check facts and information. An anonymous reviewer greatly improved the manuscript. Vincent Burke did an excellent job as the editor from concept to completion. I very much appreciate his insight, encouragement, and hard work. The content and opinions are mine, as are the sins of commission and omission.

SAVING SEA TURTLES

1

SEA TURTLES IN THE MODERN WORLD: WHERE DO WE STAND TODAY?

Everyone with a heart appreciates sea turtles. Some of us have bought stuffed green turtle toys for our children, others display sea turtle carvings made of stone or wood, and the most fortunate of us visit places like Hawaii, Cancún, and the Cayman Islands to watch sea turtles in the water. I have devoted the better part of my life to studying these impressive and, in their own way, beautiful animals. I invite you to come with me as I journey through my memories of the world of the sea turtle. We will visit them as helpless eggs and determined hatchlings, as adventurous juveniles and wise adults. We will look at how sea turtles lived in the past and the challenges that they face today.

The story of sea turtles is also a story about humanity and the choices we make. These marine reptiles, which have been on this planet since long before the first humans arrived, are much too close to the edge of extinction. They will not disappear today or tomorrow, but if things continue as they are, some species of sea turtle will probably become extinct before the end of this century. Yet the situation is not hopeless, and indeed—as I hope to reveal—the threat of extinction shares the stage with optimism. No matter what the problems are, I believe that we can make this a better world for sea turtles.

They can be saved, I like to tell audiences when I finish a talk, if we *choose* to save them.

This journey, narrated by me, is not meant to be about me. And so I will introduce you not only to the several species of sea turtles but also to some of the people who study and are working to save sea turtles. Sea turtles are doing better in some places than they were 50 years ago because of the efforts of these sometimes brave, always devoted, scientists and conservationists. In other places the story is bleak, as the right person has not yet arrived on the scene and the turtles are suffering at the hands of destructive fishing methods, overdeveloped beaches, or the direct slaughter of adults and nests.

Scientists, government officials, professional conservationists, volunteers, students, and concerned citizens form a loosely affiliated army trying to hold a line of protection around the remaining sea turtle populations. In many areas, the line has been held and built stronger because of the dedication of one person. The cliché that one person can make a difference—that you can be that person—turns out, in this case, to be true. Come and see what others have already done and why so much work remains.

Why Are Sea Turtles in Trouble?

People kill sea turtles to make jewelry and drums from their shells, to eat their meat and eggs, and because turtles get in the way as houses and hotels are built on top of their nesting beaches. Some people kill turtles by accident while harvesting the ocean's declining bounty of fish or dragging nets through the water in search of shrimp and other delicacies, or as they dredge channels for our ever-larger ships. Simply put, sea turtles are in trouble.

In 2010 everyone in the United States learned that their own lifestyles were having a direct impact on sea turtles. The Great BP Oil Spill filled the Gulf of Mexico with oil and chemical dispersants. All of the plankton, fish larvae, and jellyfish in the spill area died. Dr. Blair Witherington of the Florida Fish and Wildlife Conservation Commission e-mailed me to say that the oil sheen was everywhere but that the thickest oil was lined up in the convergence zones, where ocean currents came together. That is where all the crabs, other invertebrates, and small fish concentrate along with the *Sargassum* and other seaweeds. That is where the juvenile sea turtles hang out. We watched on the news as a few pelicans and a couple of Kemp's ridleys were released after being laboriously cleaned of oil. The news did not show us the thousands of sea turtles torched in the gulf as oil company contractors burned off large areas of oil on the sea surface. Unbelievable as it may seem, they made no effort to pick up those turtles and save them. They just lit the fires and burned them alive. They even stopped boats that were trying to save the turtles. That was the real cost of the oil to heat our homes and the gasoline to fuel our SUVs. All of the effort to save all of those Kemp's ridley eggs and hatchlings at Rancho Nuevo in Mexico and all of the regulations to avoid drowning turtles in shrimp nets went for naught. Ten year classes of Kemp's ridleys, the most endangered sea turtles, and loggerhead turtles from the most threatened populations that nest along the Gulf coast, oiled, starved, and broiled alive. Years of conservation efforts gone up in smoke and a vast area of ocean made into a desert for turtles, fish, and other wildlife. This will affect sea turtles in the Gulf of Mexico for a generation, both human and turtle.

Each one of the threats sea turtles face connects to all the others because most species of sea turtles face all the threats. And each has its own solution. One of my vegetarian conservationist friends once told my colleague Frank Paladino that the problem was simple: people eat too much meat. By "meat" she meant

the muscles of animals—whether fish, or shrimp, or clams, or turtles. I suppose that some people think of sea turtles as useful animals like cows or deer. I have even heard people say that God put sea turtles here for us to use as we see fit, even to the last animal. Most people, I think, disagree with that sentiment but may not consider the effect their actions have nonetheless. Someone wants a house on a beach, and that is understandable. But that house, with its bright lights, disturbs the mother turtle when she comes ashore to nest. Even if she doesn't abandon her attempt to dig a nest, those lights disorient the hatchlings that eventually emerge. Instead of heading to the ocean, where the light is "supposed to be brightest at night" they head to the highways and die. In other places maybe people don't realize that taking the sand off the beach to make mortar and cement will diminish a turtle's ability to lay eggs on that strand.

And yes, some people, like the Grinch in Dr. Seuss's Christmas story, have hearts that are two sizes too small. Their hearts don't pump enough "loving blood," so the brain shrinks and the neurons shrivel until the only thing that fits inside is the $ symbol. All they care about is the money that they are going to make. Catching more fish, building more houses and hotels, or selling more eggs simply means that $ becomes $$ becomes $$$. I like to think that God put us here, not to starve, but to care for the world while treating each other, and all of the earth's treasures, with respect.

The Trouble Begins with Eggs

Many egg collectors—typically called poachers, or *hueveros* in Spanish—like some fishermen and most developers of turtle beaches, have an undersized heart. *Hueveros* deal in smaller amounts of money than the developers, and it is true that they need to keep the mother turtles alive to collect their eggs. But in today's world, with all the conservation information available, many poachers know they are breaking laws and doing harm. They just don't care about the consequences of their actions. Other *hueveros* believe they are continuing a noble tradition and won't accept that doing what their grandfathers and fathers did before them doesn't work anymore. In heavily populated areas, the *hueveros* sometimes take all the eggs on the beach. One or two human generations of overharvesting and the tradition will come to an end anyway.

In January 2009, I was on Playa Nombre de Jesús in northwest Costa Rica with my graduate student Gabriela Blanco. Gabriela was studying the reproductive biology and migration of green turtles (*Chelonia mydas*), specifically eastern Pacific green turtles. When a turtle was well within the trance-like state it enters when laying eggs, Gabi would use a portable ultrasound machine to look at the turtle's ovaries (the same kind of ultrasound that doctors use to examine developing fetuses in women). Gabi would attach a satellite transmitter to the shell, and the turtle was then free to return to the ocean. She could then record its movements as it migrated back to its foraging grounds. Because modern transmitters also measure more than movement (diving depth, for example), Gabi could also determine a few things about how the turtles lived at sea.

Gabi is from Argentina. She's a proud, brave, and beautiful woman. Like pretty much every woman who devotes herself to studying or conserving sea turtles, Gabi is tough. But none of that prevented her from being thwarted by poachers. A year earlier, when Gabi began her research, I thought she would be fine, because Rotney Piedra, director of nearby Parque Nacional Marino Las Baulas, and his wife, Elizabeth, had been carrying out research on the green turtles on these same beaches for several years. They always brought rangers over from the park, and the result was a dramatic decline in poaching. However, Rotney

 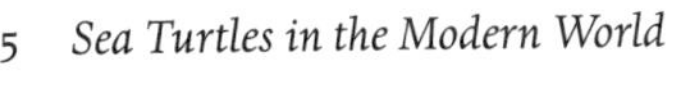

was having trouble staffing for all the urgent needs he had to manage. There was a shortage of personnel, and Gabi had no ranger to back her up on many nights. Word got around. Soon the beach was overrun with poachers, who took every clutch of eggs laid by the turtles. Unfortunately, this is common on beaches in Costa Rica and other Central American countries. If there are no guards on the beach, most or all of the eggs are poached. Brave as she was, Gabi was no match for groups of men who were intent on getting her out of the way so they could reap their bounty.

Gabi's first year of efforts proved to be mostly a waste of time. She would attach transmitters to turtles only to find out later that poachers had removed them when the females came back to the beach to lay more clutches of eggs. Gabi spent some of her time that year tracking poachers' trucks as they carried transmitters away from the beach. One transmitter stayed on the move. It traveled about 30 kilometers (18 miles) to Belen. Gabi visited various houses in Belen asking for her transmitter, but no one, of course, knew anything. She then checked her computer and saw that the satellite reported the transmitter to be at the Liberia Costa Rica airport. Soon it was farther away, in Puntarenas, the main west coast port. Then the transmitter stopped transmitting. I wonder if the turtle was still attached when the transmitter broke or was destroyed.

In November and December 2008, Gabi brought along her own reinforcements. They were two of my former students, Stephen Morreale, now on the faculty at Cornell University, and Frank Paladino, now a professor at Purdue University. Rotney also sent some rangers from Las Baulas Park. Together they chased away the poachers. By January 2009, where I began this tale, there was only one poacher left. Someone joked, "It's safe enough for Dr. Spotila." So I flew down, Gabi wanting to show me how her work was going.

The lone nighttime poacher—his name was Ricardo—walked the beach somewhat drunk and scaring away the turtles with his white floodlight. His presence was hampering Gabi's efforts. Another of my graduate students, Carlos Mario Orrego, a Tico (that is what Costa Ricans call themselves) simply talked to Ricardo. They seemed to come to an understanding. We'd turn a blind eye to this last, seemingly immovable poacher if he didn't bother Gabi and her turtles. So that night we got to work, and for the rest of the season things went pretty well. Gabi was able to make some real progress. In fact, Ricardo even found turtles for her. At one point he held her red flashlight while she did her ultrasound work, and in general left her and the turtles alone. He once told Gabi that he preferred to be called Richard because he liked the way the gringo name sounded.

"Richard" and Gabi sort of became friends. He still put his knife on the bag of eggs he collected so Gabi would not steal them. She asked him why he took the eggs, and he said it was to feed his family. Then she asked him how many clutches he took each night. He said, "Well, usually five," because that was all he could carry. Five clutches of a hundred green turtle eggs a night! I guess he had a big family—and perhaps an undersized heart. Of course we hope that his heart is growing, just as the Grinch's did. Maybe in future seasons Ricardo will become a protector of the turtles. That has happened in the past in some of Costa Rica's parks. For example, Doña Esperanza Rodríguez of Playa Grande was in charge of the poachers in 1988. She assigned each person a 100-m (109-yard) stretch of beach, and he or she could take all of the clutches of eggs laid in that section. By organizing the poaching, Doña Esperanza reduced the number of machete fights over who got the eggs. One day she met Doña María Teresa Koberg and her heart grew two sizes. Doña María Teresa is an amazing woman who almost single-handedly stopped the poaching on Playa Grande and got

the government to create a park there in 1990. She told Doña Esperanza about how special the turtles were and hired her to protect the nests on the beach. By 1992 Doña Esperanza helped to convert all the poachers there into ecotourism guides. Good things do happen.

A Night in the Forest

A secluded beach in Costa Rica can be a bit like the Wild West. It was a Sunday night, January 18, 2009, and Gabi, Jeff Seminoff, and I had been looking for nesting turtles. Jeff is a sea turtle biologist with the U.S. National Marine Fisheries Service in La Jolla, California. As the nesting slowed down, we saw a light on the beach. A guide was leading two children to look for nesting turtles. Well, that was it for turtles for the night. We knew they wouldn't come ashore with that floodlight turning night into day. The guide was either untrained or poorly trained. Either way the result was the same: the turtles were scared off.

We soon gave up our search and started driving our *cuatro por cuatro* (4×4) pickup truck back to the research station. We drove up a hill, around a tricky corner, and over a gully in the road to reach the top of the forested area. We were just congratulating Gabi on her great driving when we turned a corner and saw some people sitting beside the road. At first we thought they were poachers; who else would be out there at 2 a.m.? Gabi, alarmed, told us she thought they were not locals. We soon realized she was right. It was a European woman with her parents and her one-year-old daughter. The mother's parents were exhausted and waiting for rescue. They thought we were poachers or bandits.

As we talked to them we learned that they were connected to the flashlight on the beach. The guide had sent them and the woman's husband, a Tico, back over the small mountain between Nombre de Jesús and the next beach, Zapotillal, to find the car because they were too tired to keep up with him. The guide took the two older children, eight and ten years old, to see a turtle, but he misled them regarding how far away it was to the car and about how hard the hike was. When the older folks couldn't walk anymore they had to rest, so the husband went on in search of the car. The mother was from the Netherlands and her elderly Dutch parents were really exhausted and scared of the wilds of Costa Rica.

Not knowing when or if the husband would find the car, we offered them a ride, saying we would head to where their car was and also look for the husband. They quickly accepted, so we piled them into the two seats of our pickup, with Jeff and me in the back cargo bed. When we reached the gate at the main road, Jeff and I got out. We stayed there, in front of the farm that included the beach we were searching, in case the father came by while Gabi drove the family to their car. She reached it just as the father arrived and the family was to be reunited—once the guide came back with the other children!

Jeff and I were not so fine. The farm's watchman woke up and soon his rifle was trained on us. Jeff, who speaks fluent Spanish, walked over to explain to the nervous man that we were with Gabi and that we were biologists, not robbers. The man calmed down and put away the rifle. All was well for the night. It was a good thing that Jeff was there. There was no telling what would have happened had we relied on my Spanish.

Pacific Greens

Jeff had better nights during his visit. He said that the beaches we were walking supported one of the largest nesting aggregations of green turtles in the Pacific Ocean. Jeff had led a global assessment in 2002 on the status of the green turtle and knows more about green turtles in the Pacific than anyone.

The big problem for the green turtles on the beaches in Pacific Costa Rica is poachers. Sure, we chased off some poachers at night, but many simply changed their hours and came at dawn, or midday, or later. We came to believe that the poachers still dug up virtually every nest and took all the eggs—sometimes even the clutches that Gabi collected and moved to new "secret" locations. At the current rate of poaching, it will not be very long before there are no more green turtles on the beaches that impressed Jeff so much. To make matters worse, many green turtles are also being caught by fishermen in the nearby waters.

Local fishermen set longlines right along the front of the beach. These are, as the name implies, miles of monofilament line with baited hooks. You can find longlines in the Gulf of Papagayo with green turtles and olive ridley turtles (*Lepidochelys olivacea*) caught on them, two or even four on a line. These artisanal fishermen come out in small open boats and set longlines and also nets and leave them overnight. The green turtles have to swim through the nets and past the longlines when they leave the beach after laying eggs. There are ways to catch fish without harming turtles, or at least minimizing the harm. But the drive to solve these problems is not, at present, strong enough to force a change.

Global Warming

The earth is getting warmer. The process is slow, but there truly is no doubt about this well-documented fact. It has been, on average, warming since the end of the last glacial period some 10,000 years ago. There have been some cool periods, most famously the Little Ice Age in Europe from the 1400s to the 1800s, but in general the earth has been warming and seas have been rising for thousands of years. For the past 100 or so years the situation has intensified; some would say it has gotten out of control. What exactly will happen as the earth warms further is a topic of debate. Almost certainly the oceans will rise and in the process sea turtles will be endangered in a new way. The few nesting beaches left are likely to be threatened with flooding. In addition, sea turtles have temperature-dependent sex determination (they don't have X and Y chromosomes like humans). The temperature of the egg determines the sex of the developing embryo; warm temperatures produce female hatchlings and cool temperatures produce male hatchlings. Some beaches produce mostly female hatchlings today, and that may only increase as beaches heat up from global warming. No boys, no good.

Heroes

There are true human heroes in this story. Peter Pritchard was named a "Hero of the Planet" by *Time* magazine for his tireless efforts to save sea turtles. His scientific and popular writings, films, and speaking engagements have brought sea turtles to the forefront of public consciousness. Randall Arauz has worked tirelessly in Costa Rica to save sea turtles and sharks from longlines, gill nets, and shrimp trawls. When he was a student he introduced me to the sea turtles on the Pacific coast of Costa Rica. As the leader of PRETOMA (Programa Restauración de Tortugas Marinas), a nonprofit organization, he continues to "encourage" government authorities in Costa Rica to do their job, as required by law, to protect sea turtles and to enforce laws to regulate fisheries. Aliki Panagopoulou and Dimitris Margaritoulis of ARCHELON—the Sea Turtle Protection Soci-

ety of Greece—are responsible for much of the scientific and conservation program for loggerheads (*Caretta caretta*) in the Mediterranean. There are many more, and some of them will be featured in later chapters.

110 Million Years and Counting

Sea turtles have been swimming the earth's oceans for 110 million years and have, so far, survived centuries of human exploitation. Researchers Karen Bjorndal and Jeremy Jackson calculated that before humans came into the picture there were more than 100 million green turtles in the Caribbean Sea alone, and hundreds of millions in the rest of the tropical oceans. Now there are perhaps 0.1 percent of those numbers. There were more than 500,000 hawksbills (*Eretmochelys imbricata*) in the Caribbean and at least 4–5 million on reefs around the world. Now there are perhaps 1 percent of that number. I estimate that tens of millions of loggerheads swam the world's oceans before the arrival of our species. Now there are less than 2 percent of that number, at best.

Imagine prehistoric days when there were 500 million olive ridleys around the world and 300,000–500,000 Kemp's ridleys (*Lepidochelys kempii*) in the Gulf of Mexico and along the Atlantic coast of the United States. Now we are "happy" with a million olive ridleys (0.2% of the prior estimate) and 5,000 Kemp's ridleys (1%). How many flatback turtles (*Natator depressus*) were in and near Australia before people got to them? Perhaps 500,000? Now there are 50,000 adults. Leatherback turtles (*Dermochelys coriacea*) probably numbered a million at one point. Now there are only a few thousand in the entire Pacific Ocean and maybe another 30,000 adults in the Atlantic.

By any historical measure, sea turtles are in grave danger of extinction. The last few good nesting beaches are constantly under threat. If we want to restore sea turtle populations and preserve a 110-million-year heritage, here are a few rules we should live by:

1. Don't eat sea turtles or their eggs
2. Do not make leather and jewelry out of them
3. Don't build houses and hotels on turtle beaches
4. Eat less from the ocean in general and harvest what we eat more responsibly
5. Establish laws, regulations, and enforcement to ensure that oil drilling and other activities are conducted in a safe manner in the oceans
6. Increase the pressure on our leaders to advance environmental causes, including laws and regulations to halt global warming

Where Do We Go From Here?

Some, maybe all, species of sea turtles will go extinct in a century or so if we sit back and watch it happen. Archie Carr, the father of sea turtle biology and conservation, studied sea turtles and wrote about them in several books. His research revealed many secrets about their lives. His conservation efforts saved the green turtle from extinction in the Caribbean Sea and alerted the world to the perilous plight of all sea turtles, and his students carry on that tradition today. Archie did not sit back and watch it happen. He was one of the first activist scientists, a contemporary of Aldo Leopold and Rachel Carson.

Archie's students are among those who continue the fight. They are accompanied by many others. Those who join the battle to save and restore sea turtles hail from all quarters, and their efforts, if reinforced, may well win this war. The children of today will have to be the conservationists of tomorrow if

sea turtles—and all the other wild animals they represent—are to thrive. On every beach where sea turtles nest, and in every ocean where they swim and feed, the fight to save and restore them must be fought. If everyone reading the stories I am about to unfold selects one species in one location and supports that effort, we might turn the tide. Select a piece of the world—one sea turtle beach perhaps, one fishing regulation, one restoration project—and support the efforts related to it. Teach our children to make a difference by making a difference yourself. The work of conservation continues. Archie Carr's dream of sea turtles thriving (not just surviving) in the world's oceans lives on. We still have time. Will you join us?

2

LIFE IN THE EGG: BURIED ALIVE UNDER TWO FEET OF SAND

A sea turtle comes ashore and digs a hole in the sand, drops her eggs, covers them up, and leaves them to incubate for 45–60 days. The hatchlings she will probably never see eventually crawl up through the sand and boil out of the nest at night before running down the beach to reach the ocean and swim off. Sea turtles have been completing these same steps for 110 million years. It is a good system and has plenty of excess capacity, so that if natural predators such as raccoons or their cousins, the coatis, dig up some nests there are plenty of extra eggs. You could say that the turtles follow the old adage, safety in numbers.

How Eggs Work

A sea turtle embryo develops inside an eggshell that surrounds the albumen (egg white) and yolk and keeps it from drying out but allows for gas exchange during development. As the one-celled fertilized egg, called a zygote, divides into a two-, four-, eight-celled embryo and so on, it consumes oxygen and produces carbon dioxide. As the embryo develops into a larger and more complex organism, it needs more and more oxygen, produces more and more carbon dioxide, uses up most of its yolk supply, and produces waste that has the potential of poisoning it.

One of the great events in evolution occurred about 300 million years ago when a group of amphibians arose with a new kind of egg. When complete, this egg had three special membranes and a unique shell that allowed the mother to lay her eggs on land instead of in water. That amphibian mother became the first reptile, and it is those membranes that allow the sea turtle embryo to survive buried in the sand and to grow and develop into a hatchling. Getting the egg out of the water reduced danger from aquatic egg eaters, giving those early reptiles a great advantage and a permanent foothold on land. No longer did they have to return to water to lay their eggs. Reptiles expanded far and wide and eventually gave rise to dinosaurs, birds, and mammals.

The sea turtle embryo, like all reptile embryos, sits in a sac of liquid surrounded by the amnion, a membrane that develops inside the egg soon after it is laid. The amniotic liquid serves as a shock absorber and keeps the embryo bathed in clean fluid. When a male sea turtle mates with a female, some of his sperm rush to the ripe eggs and fertilize them while other sperm are stored in the oviduct for use with later, still-developing clutches of eggs. The fertilized egg, consisting of the fertilized ovum and its yolk, then descends through the oviduct, where it gets covered in albumen and then a leathery shell. The zygote divides until it reaches an eight-cell stage and then stops developing and enters a period of rest called a diapause. The pliable shell and large amount of albumen allow the egg to drop 30–80 centimeters (12–31 inches) from the mother turtle to the bottom of the egg chamber without injuring the resting eight-celled embryo.

Once the egg is laid in the nest, it begins to develop again. After about 8–12 hours, the yolk rotates inside the egg, putting the young embryo at the top of the egg and near the inside of the shell. Then the embryo starts to adhere to the shell membrane.

As this happens, a small white spot appears on the eggshell and gradually grows to cover the entire shell as it changes from a wet, translucent cream color to a more parchment-like white. The embryo then develops the amnion and another membrane, the chorion, which will be filled with blood vessels that connect to the embryo and help transfer oxygen and carbon dioxide between the embryo and shell.

Another membrane, the allantois, also develops around the embryo as an outgrowth of the rear portion of the embryo's digestive tract, or gut. It forms a sac that receives waste products of metabolism from the embryo. These include nitrogen compounds, such as ammonia and urea from breakdown of the yolk, which are poisonous and would kill the embryo if left in contact with it. As the embryo grows and produces more waste, the allantois also grows larger. Later, the chorion and allantois merge to form a combined membrane, the chorioallantois, which serves as the embryo's means of taking up oxygen and getting rid of carbon dioxide during much of its incubation period.

The yolk provides fats and other nutrients such as sugar, starches, proteins, and vitamins used in development, and the albumen provides lots of water and some proteins. The many blood vessels in the yolk sac carry nutrients to the developing embryo. As the embryo grows, yolk is used up and the yolk sac shrinks. Space inside the eggshell is taken up by the enlarging embryo. When a sea turtle hatches, the yolk sac may already be so small that it is absorbed into the body of the hatchling. However, often the hatchling has to lie in the broken shell for a day or two to complete the absorption of the yolk into the body cavity. The hatchling cannot safely leave the nest if the yolk is still hanging out of its body, because a ruptured sack would kill the hatchling.

Sea turtles have flexible eggshells, not the more familiar

hard-shelled variety seen in birds. Because the turtle eggshell is thin and porous, gases pass through it easily. An oxygen molecule from above the ground moves passively through air spaces around the sand in the beach down to the nest. Then it passes through the eggshell and enters blood vessels inside the chorioallantoic membrane. Red blood cells absorb the oxygen molecule and carry it to the embryo.

Because the embryo is submerged in amniotic fluid and its lungs do not yet work, oxygen is carried directly to the cells of the developing embryo and used to fuel its metabolism. Nutrients in the yolk are turned into turtle. Carbon dioxide is taken up by red blood cells and the journey is reversed until the carbon dioxide molecule reaches the free air above the sand. The water keeps the embryo hydrated and the excess passes readily through the eggshell into the surrounding sand. The other waste products go into the allantoic sac. When the egg is laid, the shell is dimpled because the egg is not fully hydrated. This provides some flexibility to the shell so that it does not split open when it falls into the nest chamber. Within a few days the egg absorbs water from the sand and becomes perfectly round.

Buried under the Sand

There are advantages and disadvantages for burying sea turtle eggs under the sand. Two big advantages are that the eggs are less accessible to predators and they incubate at a more stable temperature than if they were above ground. Those advantages vary from species to species. Olive ridley and Kemp's ridley nests are only 38 cm (15 inches) deep at the top, so they can be found and dug up fairly easily by dogs, coatis, vultures, and other predators. Maybe that is why olive and Kemp's ridleys use the *arribada* nesting system, in which hundreds of thousands of eggs are laid simultaneously by thousands of mothers in a couple of days—the predators can only eat so many. Green turtles and loggerheads bury their nests deeper, the top of the eggs lying about 50 cm (20 inches) down. Leatherbacks put their nests deeper still, with the top of the nest about 60 cm (25 inches) or more from the surface. The deeper the eggs, the more stable the temperature. At a depth of 20 cm the sand temperature can vary about 2°C over 24 hours, while at 50 cm this is reduced to less than 1°C. At 70 cm the variation is less than 0.5°C.

On the other hand, there are two big disadvantages to burying eggs deep within the beach. First, if the nest is too deep or too close to the water, eggs can be drowned by the water table as the tide comes in. Second, eggs have to breathe through all that sand. There is water under the beach and the depth of that water varies with the amount of rain that the beach receives and with the height of the tide. During the rainy season in the tropics the beach can be inundated with water and eggs can drown. Leatherbacks, therefore, tend not to nest in the rainiest time of the year because their eggs are so deep in the sand. Green turtles can nest during rainy periods because their eggs are shallower, especially in areas like the Caribbean, where tides are minimal.

Ridleys lay their eggs in the rainy season because in the dry season the sand begins to dry out and a distinct dry front develops in the sand, whereby the upper level of sand is as dry as a desert and the lower sand is moist. There is no gradient in moisture; it is either wet or dry. As the dry season progresses, this drying front sinks deeper and deeper in the beach. Ridley nests are shallow enough that the dry front intercepts the nest and the eggs dry out. So those ridleys that lay their eggs in the dry season don't produce any hatchlings. That is a good example of how evolution works. The environment selects against those ridleys and they do not pass on their genes to the next generation.

It is not easy to breathe when you are buried under 25–50 cm

(1–2 feet) of sand. That is true for people but less so for sea turtle embryos. In fact, sea turtle embryos seem to do quite well in getting O_2 and getting rid of CO_2 while buried under the sand. It was not until the 1970s that someone began to figure out how those embryos did it. Ralph Ackerman, a thoughtful person who is at home with both physics and biology, was a graduate student at the University of Florida, where he studied gas exchange in green and loggerhead turtle eggs and calculated the heat and mass balance of the eggs. He carried out clever experiments in the laboratory to determine the rates of diffusion of gases and water through the sand and into and out of the egg. He also measured the O_2 and CO_2 exchange between the egg and sand. These groundbreaking experiments are the basis for our understanding of how the developing eggs work in nature.

Ackerman worked in Florida and Tortuguero, Costa Rica, where the difference between low and high tide was much less than 1 m (3 feet). In the 1990s and 2000s we were working at Playa Grande in Las Baulas Park on the Pacific coast of Costa Rica, where tidal range was 3 m (more than 9 feet). We were measuring the O_2 and CO_2 exchange and temperature effects in leatherback eggs. One of our team members was Paul Sotherland, a professor at Kalamazoo College. Paul is a tall, lean, and energetic guy with an infectious grin and wind-blown hair. He was walking on the beach during low tide in 2003, when he noticed water draining out of the sand below the high tide line above the edge of the ocean. He wondered if the tide going in and out pushed up the water table beneath the sand. Did that big tidal flux affect the concentrations of O_2 and CO_2 that we were measuring in leatherback nests in the hatchery and in natural nests in the beach? It seemed that the values of O_2 were higher and CO_2 were lower than predicted from Ackerman's equations. Paul hypothesized that the tide acted as a bellows and pushed air up and down in the sand, ventilating the nest. The problem was how could he measure that effect? Did the water table under the beach actually move up and down with the tide?

Theoretically all you had to do was to dig down to the water table and measure it. So that is what Paul did—he started to dig a hole. He began early in the morning, before the sun was up, so that he could finish before it got too hot, and by late afternoon we wondered how he was doing. We walked down to the beach and there he was, in a hole about 3 feet wide and as deep as he was tall. The reason he was still there was that he was stuck in sand, buried past his knees because the walls of his hole had collapsed. After a short discussion about safety we convinced him that new measures were needed.

One option was to have the students dig the hole. There were more of them than of us professors. Paul didn't like that idea—something about losing their tuition dollars if the hole collapsed again. So we compromised. A couple of students volunteered to work with Paul, and we got him a ladder. They dug the hole wider than it was deep and kept making it wider at the top as they went down. The hole was terraced so that there was no more than 3 feet of vertical surface in the hole before a nice flat spot and the sides would not bury the workers if the sand fell in. The ladder was great because Paul and the students could get out of the hole!

By the next day they hit water. It was more than 10 feet down. Next Paul rigged up a 6-inch-wide piece of plastic PVC pipe and set it in the hole so the bottom was in the water. Then they back-filled the hole. By placing a toilet ball float in the pipe with a rod attached, Paul could watch the float go up and down with the tide. For the next few weeks the students sat by the pipe and recorded the level on the rod every hour, and thus captured the tidal cycle and change in depth of the water table under the beach.

The beach breathes! As the tide comes in and out, the water

table goes up and down and like a bellows it pumps air past the turtle nests. That increases gas exchange, so that levels of O_2 are higher and levels of CO_2 are lower at Playa Grande than in sea turtle nests on Caribbean and Florida beaches. Paul was right; all he needed to test his idea was a shovel, a ladder, a toilet ball float, and some dedicated students.

Temperature and Development

Temperature controls the rate of development of a sea turtle egg. As in all cellular processes, the higher the temperature, the faster the metabolic rate. So if a sea turtle digs her nest in the open beach, the clutch of eggs will be warm because the sun beats down on the beach all day. In the shade the sand will be cooler. If it rains the sand will also be cooler, even in the open beach, and incubation will take longer. If the sand is cool, say 25°C (77°F), incubation can take up to 85 days. In the open beach eggs can reach 35°C (93°F) and eggs will hatch in fewer than 45 days. Add to this the effect of metabolic heating and in a large clutch of eggs temperatures can rise 3–5°C above sand temperature in the last week or two of incubation.

The story of how temperature works in sea turtle eggs seems pretty straightforward now, but it took a lot of hard work to figure it out. Among the first to make progress was John Hendrickson, who worked in the Turtle Islands of the South China Sea in the 1950s. He set up hatcheries on Sarawak Island to help save green turtle eggs from local people who overexploited the eggs. In the process he noticed that in the rainy season eggs took longer to hatch. He drew a graph that showed that the more rain that fell, the longer the incubation period. He figured this was related to temperature, but he was unable to collect much detailed data on temperature because he did not have the sophisticated thermocouple apparatus that many sea turtle biologists use today.

In 1955 there were three ways to measure temperature in a turtle nest. One way was with a mercury thermometer. Of course, a big glass thermometer full of mercury was not easy to use on the beach. Imagine sticking a long mercury thermometer into the sand and trying to get it into a turtle nest. The thermometer is big and fragile, so that is almost impossible to do. Even more difficult would be the trick of getting it out to read the temperature before it changed on the way up through the sand. Then you would have to repeat that process for every nest each day that you wanted to make measurements.

The second method was to use a less sophisticated thermocouple than we use today. A thermocouple is made up of two different metals that generate an electric current when they touch. You may have experienced that electric signal if you ever bit down on a piece of aluminum foil and it touched a filling in your tooth. That is certainly a shocking experience and I do not recommend trying it, but it proves that two different metals do generate an electric current when they touch.

Thermocouples were available at that time, but they were almost impossible to use on the beach. First of all, the thermocouple system of the day compared the voltage generated between a measuring junction and a reference junction. You would need something that would give a known and constant reference temperature so that you could place the reference junction there and then compare it to the measuring junction. That reference temperature was typically provided by an ice bath, notoriously absent on a sea turtle nesting beach. If you put a mercury thermometer into the ice bath you would know its temperature, which was not always exactly 0°C. Then you would need a laboratory meter to read the difference in mil-

livoltage between the two junctions. That was okay as long as you had electricity to run the meter, which Hendrickson and his contemporaries did not.

In the 1960s we made these measurements in the field by obtaining a battery-operated, portable millivolt meter. It was awkward to use and you had to change out the circular measuring dials as the temperature changed from near 68°F to about 86°F. As long as your ice lasted and your thermometer did not break, you were good to go. Of course, you would have to set up a measuring and reference junction for every nest that you measured, so thermocouples were not a practical alternative.

The third method involved a thermistor, a variable resistor, which is made up of a metallic oxide that changes electrical resistance with temperature. If you have a thermistor that changes resistance linearly with temperature and you hook it up in a circuit called a Wheatstone bridge, which compares the resistance of the thermistor with three other resistors, you can use an ammeter to read the change in current in the bridge circuit as the thermistor changes resistance with a change in temperature. Calibrate the change in current with resistance change of the thermistor versus temperature and you can convert the current reading to temperature. That is all there is to it.

I love the name "Wheatstone bridge," and that was one of the few things that actually worked for me in physics lab at the University of Dayton. However, thermistors were not linear in 1955 and it was very difficult to get a portable system into the field that actually worked. In the 1960s the Yellowstone Instrument Company solved those problems and produced the YSI telethermometer, a versatile field-portable thermistor thermometer with waterproof probes that you could place in the field and come back to measure at any time. But in the 1950s there was no such option.

So what did Hendrickson do? He was very clever. He figured out a way to use laboratory thermometers made of glass and mercury to measure the temperatures of the sand and of clutches of turtle eggs. He hollowed out bamboo shoots and buried them in the sand so that some reached into the turtle nest and some were in the sand a meter or so nearby at nest depth. Then he placed thermometers into the bamboo holders and let them sit for an hour or more with a cork at the top to avoid air movement.

Bamboo, like all wood, is a great insulator. So the tip of the thermometer measured only the temperature where it was placed. By tying a piece of string onto the thermometer he was able to pull it up and read the temperature at any time. Hendrickson also took advantage of the bulky size of the thermometer to avoid temperature errors: He determined that if he pulled the thermometer up quickly he could read the temperature before it had time to change. What an amazing achievement! Using this method, Hendrickson was the first to discover that the temperature was higher in a clutch of sea turtle eggs than in the sand nearby (this is known as the phenomenon of metabolic heating in sea turtle nests).

Today we use a variety of handy, convenient thermocouple meters to measure the temperature in sea turtle nests, in the beach sand and in the air. After the development of the integrated circuit by the National Air and Space Administration (NASA) for the U.S. space program, someone figured out that you could use an integrated circuit to generate a reference millivoltage right in the meter. Then all you had to do was to make a measuring lead by twisting together the uninsulated ends of two wires, usually copper and constantan, soldering the measuring junction so it would not come apart, and attaching the other end to a connector that would fit into a receptacle in the meter. Now we make thermocouples for less than $5 each and make measurements all over the beach.

The higher the temperature in the nest, the faster the embryos develop. If the temperature is 26°C (78.8°F), green and loggerhead turtle hatchlings emerge in 80 days. At 30°C (86°F) they emerge in 54 days, and at 32°C (89.6°F) they emerge in 46 days. That includes the time it takes to crawl up out of the nest, so you have to add a day or two to those values for leatherback hatchlings, which have farther to crawl to reach the surface.

Rain can make a big difference in nest temperature. Georgita Ruiz found that a big rainstorm at Playa Nancite dropped the temperature of olive ridley nests from 33°C to 30°C (91.4°F to 86°F) in 5 days and it took another 5 days for the temperature to recover to 32°C (89.6°F). Four days of rain at Tortuguero lowered sand temperature at green turtle nest depth (50 cm) from 31°C to 29°C (87.8°F to 84.2°F). The effect was less because deeper sand buffered the effect of cold rain. During the same period, at ridley nest depth the temperature in Tortuguero sand dropped from 31.5°C to 27.5°C (88.7°F to 81.5°F) because it was less buffered.

Two other factors affect nest temperature. Clutches laid below the high tide line usually die by drowning. However, eggs can tolerate tidal flooding for a few days under some circumstances, although it greatly affects their temperature. An olive ridley nest inundated by the tide for several days at Playa Nancite dropped from 32°C to 26.5°C (89.6°F to 79.7°F). That affected the sex of the hatchlings because the drop occurred during the temperature-sensitive middle third of development (see below). Only 10 percent of hatchlings from that nest were female.

Metabolic heating can also have a big effect on the temperature of a clutch of eggs near the end of development, but it has little effect on development rate and usually does not affect sex determination. It does, however, give us a signal of when a clutch is about to hatch and emerge.

Temperature-dependent Sex Determination

Temperature not only controls the rate of development of sea turtle embryos, it also determines their sex, a process known as temperature-dependent sex determination (TSD). When I was giving a talk to children in the Bullis Charter School in Los Altos, California, a couple of years ago, I asked if anyone knew how a sea turtle hatchling became a boy or girl. I was surprised when a sixth grader said that the temperature of the egg made it a male or female because the turtle had no sex chromosomes. Those were some really smart children, and also some really good teachers. But such information is not always child's play. It was a striking discovery when my students and I learned how sex determination worked in green turtles at Tortuguero, and the implications for conservation were far-reaching.

The starting point for our discovery was actually 1972. Frenchman Claude Pieau discovered that the temperature of incubation affected the sex of several freshwater turtles and tortoises. This astounding discovery received little notice in the United States. Then in 1976, Chester Yntema reported the same effect for snapping turtles (*Chelydra serpentina*) in New York. At that point Pieau's discovery started to receive the attention it deserved. By the late 1970s, TSD was creating quite a buzz and scientists were starting to wonder if this phenomenon occurred in all turtles (it doesn't). At about this time, Ed Standora and I wondered if temperature determined the sex of sea turtle hatchlings. If it did, the phenomenon would be important in conservation programs around the world in which sea turtle eggs were moved to hatcheries to protect them from disturbances. They might, for example, be producing all males. We talked to Archie Carr, and he suggested that we go down to Tortuguero and conduct an experiment to test those ideas.

We learned that Jack Woody, director of the Sea Turtle Pro-

gram for the U.S. Fish and Wildlife Service (USFWS), was looking for someone to carry out just such a study. Jack was concerned because there was a big program underway to save the Kemp's ridley turtle in Rancho Nuevo, Mexico, which involved moving eggs to hatcheries along the beach to increase their survival. What if they were producing male hatchlings? He was already talking to David Ehrenfeld at Rutgers University about these ideas. David was a leading sea turtle scientist, having worked out some of the important mechanisms whereby hatchling sea turtles found their way from the nest to the ocean. He was also a great philosopher of science and a leading conservationist. At Archie's suggestion we got together with David and his team and we set out to investigate this phenomenon with funding from USFWS through Jack Woody.

Ed and I were at the State University College at Buffalo, New York, and David was at Rutgers University in New Brunswick, New Jersey. That meant a road trip! Ed and I drove down, and David and his wife Joan very kindly hosted us for a lunch at their home. There we met Georgita Ruiz, a veterinarian from Mexico, and Merry Camhi, David's graduate student. We brought in Stephen Morreale, a former undergrad in my lab who was going to tackle this problem for his master's thesis. That afternoon we presented our experimental ideas and we worked out the details of the plan. In the summer, we all set off for Costa Rica.

At Tortuguero David suggested that we set up a model hatchery, so we began to put thermocouples into nests on the beach and set it up. The hatchery worked very well. Based on the many years of experience of Archie Carr and his students, we placed the clutches of turtle eggs about a meter apart and put a small hardware cloth screen around each nest to catch hatchlings when they emerged. The entire hatchery was surrounded by a taller fence to keep out coatis and dogs. We shaded part of the hatchery with palm thatching to make it quite cool, another area was partial thatch to simulate partial shade, and in a third area we buried eggs only half the normal depth to see if the greater temperature fluctuations at shallower depth were important. The fourth portion of the hatchery held clutches of eggs buried in the open sun at normal depth. Georgita and Merry worked with us to learn the techniques and how to make the temperature measurements. Then they went off to Playa Nancite on the Pacific coast of Costa Rica to replicate the study on an olive ridley *arribada* beach. Georgita and Merry were measuring the effect of temperature on olive ridleys because there were too few Kemp's ridleys in Mexico to manipulate their nests. The women set up a hatchery as we did and set up a camp to make measurements for several months.

The number one rule in fieldwork is that you should expect the unexpected. We had some logistical problems. The female turtles did not pay much attention to the fencing that surrounded the natural nests and destroyed it as they tried to nest. We ended up piling up big logs, driftwood, old mooring buoys, and anything else that washed up on the beach to keep the adult turtles from crashing through our hatchery and field nests. Then the rains came. It rained every day. It seemed our clothes never dried out. Towels remained moist, and our gear started to suffer from the high humidity. If you stopped moving for more than a few minutes, moss would grow on your back—or so it seemed. We solved part of the problem by setting up a dry box in one of the footlocker trunks with which we transported our gear. We put a light bulb inside the trunk and left it on to drive out the moisture. That kept the instruments, film, and paperwork dry. The light was on only for the 4 hours each day that the generator was running, but that was enough to do the trick.

By the end of the season we completed temperature mea-

surements of the beach and nests, collected some of the hatchlings and removed their gonads, and headed home. Unfortunately there was no way to conduct the study without killing some hatchlings. We had to remove the gonads and take them back to the laboratory to cut the tissue into thin slices and mount them onto microscope slides, because the only way to tell the sex of a sea turtle hatchling is from observing the microscopic structure of the gonad. We humanely sacrificed the hatchlings so that they felt no pain. It was sad to put an end to the baby turtles, both because they were so cute and because as conservationists we wanted to produce more, not fewer, hatchlings. We took special care to do statistical power analyses of our experimental design to be sure that our sample sizes were large enough to produce statistically significant results, but not larger than needed. It would be a tragedy to conduct those experiments and find out that our sample sizes were too small and thus the hatchlings that we sacrificed died in vain. It would also be a tragedy to sacrifice more hatchlings than were needed to get the data to give a clear answer to the question of whether temperature affected their sex. We also did sand-table exercises to practice how to place thermocouples and how to sample the clutches in the nests.

You have to be both a good scientist and a good person to do the right thing for your experimental animals and for conservation. Every colleague and student that has worked with me on sex determination or toxicology studies of turtles has remarked that it is not a pleasant thing to sacrifice a turtle embryo or hatchling. We took no pleasure in doing so. Every university has an animal care committee to be sure that researchers treat their experimental animals humanely. It is a special burden for field biologists far away from campus to be sure that their animals are treated well, because no one is looking over their shoulder.

In fact, we always saved more sea turtle hatchlings than we

sacrificed, because we took eggs from nests laid below the high tide line. These "doomed" clutches would have been lost anyway. We also moved additional clutches to a hatchery or a safe place on the beach so that they would survive, and we patrolled the beach to prevent poaching. On balance, more turtles survived because of our studies than would have done so if we were not there. We obtained important scientific information with a minimum of turtles, but that did not reduce the personal burden of ending the lives of those hatchlings. That is a burden that we carry to this day. We hope that our studies save many more turtles into the future.

Meanwhile, Georgita and Merry survived food shortages, crocodiles that lived along the beach, and rainstorms that washed away their hatchery when the estuary behind the beach blew out in a big flood. They also returned with temperature data and gonads so that we could complete the project that fall.

Back in Buffalo, Georgita set up a histology lab, sectioned the gonads, made the slides and examined the cells to determine the sex of the hatchlings. When we analyzed the data—sex of turtle hatchlings versus temperature in the natural nests and hatchery nests—we discovered that, indeed, sex was determined by temperature and, yes, that could be a big problem for sea turtle conservation. In Florida, people were putting eggs in Styrofoam coolers in air-conditioned house trailers so they would be "protected" and would not overheat. Bad idea! They probably produced only male turtles. Consequently, none of them came back to nest as adults.

In parts of Southeast Asia, people were putting eggs in the sand in open sunny areas. They probably made only female hatchlings. That was not a good idea, either. At Rancho Nuevo, our later calculations indicated that early May temperatures were below the critical level to produce females and that by June temperatures in the sand were above the critical temperature

range, so those nests produced only females—unless it rained. A big rainstorm could drop the temperature in the sand by 1 or 2 degrees and change turtles that were going to be females into males. Rancho Nuevo, therefore, probably produces a female bias in the hatchlings that come off the beach.

We were pleased when we learned a couple of years later that Nicholas Mrosovsky, a biologist at the University of Toronto, had completed similar experiments on loggerhead turtle eggs in the United States at about the same time as our work in Costa Rica. His results confirmed our findings and showed that seasonal changes in sand temperatures could be important on beaches as well.

Threats on the Beach

On too many sea turtle nesting beaches most of the eggs are taken by people, with a few left for their dogs. When combined with death-dealing processes in the oceans like gill-netting and long-lining, entire sea turtle populations are wiped out. For years, people in Malaysia collected 90 percent of leatherback eggs on Terengganu beaches. There were plenty of turtles, and plenty of eggs. Then one year the population crashed. Adult females died of old age or got caught in fisheries in the South China Sea and no longer came back to nest. No new turtles replaced them so the population evaporated. Saving only 10 percent of the eggs wasn't enough.

Common sense tells us that taking most eggs of a sea turtle population would cause it to collapse. Dr. Pilar Santidrián Tomillo, who likes to be called Bibi, demonstrated this scientifically in a study of leatherback turtles. Bibi's mathematical analysis of egg poaching put the nail in the coffin of the idea that you can save a sea turtle population by protecting only a few eggs in a hatchery. Bibi is from Madrid and has dedicated her life to studying and saving sea turtles. She is a bright, creative scientist and very effective conservationist who works well both with other scientists and local people in the community.

Bibi studied the effect of egg poaching on the leatherback turtle population in Las Baulas Park in Costa Rica. At the time of her arrival I thought that the collapse of the leatherback population at Las Baulas was due primarily to deaths caused by the fishing industry. Bibi interviewed the oldest residents of the area and created a mathematical model of the leatherback population using a Leslie matrix (a discrete, age-structured model of population growth that computes the changes in a population of organisms over a period of time) and a life table (the kind of table that insurance companies use to determine life insurance rates). She then did a series of simulations of the population under natural conditions when the population was stable and with different levels of poaching as reported for the past, both with and without the death of adult turtles owing to fishing activities. Her study clearly demonstrated that egg poaching alone was responsible for most of the decline of the population. Collecting most of the eggs for a generation of turtles in the 1970s and 1980s resulted in the dramatic decline in numbers of nesting leatherbacks years later, starting in the 1990s. Las Baulas was only a step behind Terengganu.

Based on Bibi's research, we can predict with confidence that, if local fishermen keep catching turtles and egg poachers continue to raid beaches, it will not be long before the best nesting beaches for eastern Pacific leatherback turtles will be empty. However, egg collection is virtually stopped with a modest investment of guards. Bibi's mathematical model also indicates that we should see an increase in protected populations as new turtles return to nest in coming years.

Eggs or Viagra?

Today we have Viagra and other medicines that men take to improve their sexual performance. But in Latin America and other places, men still like to drop sea turtle eggs into their glass of rum or beer and chug it down. They reason that since female sea turtles lay so many eggs, eating those eggs should make a man more macho, more masculine. This always seemed to me to be a particularly twisted form of logic. If anything, I'd figure that eating all those eggs would make a guy more feminine. In any case, an egg in a glass of beer is really gross. I don't recommend it, and I have not tried it.

It is really hard to change the culture of egg eating. A few years ago a campaign in Mexico produced posters of a beautiful woman lying on a beach with sea turtle hatchlings. The poster said in Spanish: "My man doesn't have to eat sea turtle eggs!" We thought that was a great campaign so we ran off a few color copies of the poster and put it up in Kike and Yanira Chacon's Restaurant at Playa Grande in Costa Rica. The next day at dinner one of the students heard two local men talking in Spanish while they were drinking their rum and beer and enjoying the poster. One said to the other: "Yeah, I wouldn't have to eat sea turtle eggs either if my girlfriend looked like her!"

Other people take eggs and cook them for an interesting meal. Sea turtle eggs never harden like cooked chicken eggs; they stay liquid and slimy. Chicken eggs solidify because they have lots of sulfur-to-sulfur chemical bonds in their yolk and albumen (the yellow and white parts of the egg), and that makes the eggs denature, or coagulate, when they are heated to a high temperature. Sea turtle eggs have few such chemical bonds, so they do not coagulate and remain a hot slimy glob on your plate. The only thing that you can do to make them edible is to add enough flour to thicken the eggs.

I know all this about sea turtle eggs by direct experience. In 1978 at Tortuguero, we were running out of food and the supply boat from Limón was overdue. So we were really hungry. One morning we found a green turtle nest that had been washed out by the tide and waves. The eggs would never have produced hatchlings but did still look edible. So we brought a few back to Casa Verde, the famous Caribbean Conservation Corporation biological station, and Miss Juney, the cook, whipped up some breakfast. Not the best eggs I have ever tasted, but they were edible.

Sea turtle eggs in your beer are gross and fried sea turtle eggs are slimy. What are sea turtle eggs good for? Would you believe cakes and cookies? The cookies in Costa Rica are especially tasty. Archie Carr was the first to inform me that the reason Costa Rican cookies were so light and fluffy was because they were made with turtle eggs. The lack of sulfur bonds means that eggs do not coagulate when cooked, resulting in a fluffy mix, like whipping up a meringue. In Costa Rica one of the driving forces in the egg collection business was the baking industry in San José, the capital. The major baking company there used to buy all of the eggs that it could get. Truckloads of eggs made their way from beaches of the Caribbean and Pacific coasts to San José in the central highlands. There were not many chicken farms in Costa Rica in the old days, so the cheap source of sea turtle eggs was a boon to the bakery business. Today, the bakers have switched to chicken eggs.

But the cookies are still good, even though bakers do not use sea turtle eggs anymore. Why? It seems the skilled bakers found a way to keep the taste without destroying part of the environment. Human ingenuity.

3

RACE TO THE SEA: COATIS, CRABS, AND NIGHT HERONS—OH MY!

One day the embryo is ready. It will soon enter a new world, that of the hatchling. Its new life begins with a subtle announcement: detectable changes in temperature, sound, and smell. Stephen Morreale, Ed Standora, and I discovered the temperature signal in 1980 during our experiments on temperature-dependent sex determination (TSD) at Tortuguero, Costa Rica. We saw that clutches of green turtle eggs showed a rapid rise in temperature when they neared hatching. In fact, the change of temperature overnight was so great that at first we thought our thermocouples were broken.

There are often problems related to temperature readings from a thermocouple on the beach. Sometimes the wires loosen on the terminals of the thermocouple connector that is plugged into the thermocouple meter, which can cause the temperature reading to fluctuate. Sometimes the wires get twisted from all of the plugging and unplugging of the connector into the meter. If there is any bare wire exposed inside the connector, the wires can touch. When you get a strange and unexpected temperature reading, the first thing that you do is to check the connector. That means you have to take the connector apart on the beach, try not to drop the little screws holding it together, check the wires, tighten the screws holding the wires to the terminals in the connector, and put the connector back together. And as long as you are working on the connector you might as well take out your piece of steel wool and clean off the oxidation on the connector leads, to be sure the leads are getting a good contact, and brush the sand off the connector so that you don't get it into the thermocouple meter.

After rechecking we saw that the temperatures were up a degree from the previous day. The next day they would increase another two. Soon after, the temperature started to drop, and the next day out popped the hatchlings. What's happening? When the hatchlings begin to crawl out of their eggs, their metabolic rate goes up. When they are all out of their shells they start to move around in a hurry as they climb up out of the nest. The temperature in the center of the nest rises until the hatchlings crawl above the thermocouple, and then it drops. In another day or two the hatchlings reach the surface. So we learned that by watching temperature you can accurately predict when the hatchlings will emerge.

Another way you can tell the hatchlings are on their way out of the nest is to listen. I just learned this fact in 2008, when Barbara Bergwerf and Mary Alice Monroe visited our project at Playa Grande in Costa Rica. Bibi Santidrián was telling them about studying the behavior of hatchlings and Barbara said that in North Carolina, where they are volunteers on a sea turtle project, people listen to the nests to tell when the hatchlings are going to emerge. They rigged up a microphone and amplifier and placed it on the surface of the sand above the nest. You can actually hear the hatchlings as they squirm around and crawl up through the sand.

Finally, you can *smell* the hatchlings coming. I learned that trick from another volunteer. We were sitting in the hatchery at Playa Grande with some volunteers, guarding the nests and waiting for hatchlings to emerge. We knew from the tem-

peratures that a couple of nests were ready to hatch. Andrea Young from McLean, Virginia, started to notice something: a smell coming from the sand. She remarked that it must be the hatchlings coming up. Of course, being the scientist, I knew that hatchlings did not give off a smell, so I said, "No, that can't be."

A couple of minutes later she said, Yes, there was something there. So I stuck my nose down by the sand, and sure enough I could smell something, too. It smelled like freshly turned wet soil from a garden. That was amazing! In about 30 minutes the surface of the sand started to sink into a small pit as grains of sand fell down between the cracks among the hatchlings. In another few minutes we saw a head and then another, then part of a body, and finally a whole hatchling. Here they came, twenty hatchlings pouring out of the nest. Andrea really did smell them coming. You learn some really good things from volunteers because they do not have any preconceived ideas about what the turtles are going to do. Sometimes too much knowledge closes off your imagination. Volunteers help to keep us open to new ideas and observations.

Getting Out of the Nest

It is not easy to climb out of a nest 25–75 cm (10–30 inches) down in the sand. In fact, it is very unlikely that a single hatchling can get out on its own. We excavate nests 2 days after the first hatchlings emerge and we often find single hatchlings stuck in the sand partway up from the nest. Hatchlings that do not make it out on their own are usually so exhausted after we remove them from the sand that they cannot make it to the sea. We rehydrate them in some water and hope that they will last until that night, when we release them. Owing to these natural barriers, hatchlings have evolved a cooperative behavior that allows them to emerge as a group, greatly improving their chances of success and reducing their individual effort.

Archie Carr, Larry Ogren, and Harold Hirth were the first to record the emergence behavior of sea turtle hatchlings. They carried out a simple experiment; they watched the hatchlings, in a fine example of straightforward natural history observation. It is not fashionable these days for biologists to actually observe nature, because their department chairmen, deans, and provosts urge them to get large experimental grants for research that can be done in the laboratory, which brings in large amounts of overhead money to the university. In the good old days it was possible to observe animals and not worry about overhead dollars.

Archie and company dug down to the side of a nest and put a pane of glass up against the eggs. They made the hole large enough that a person could get down into it and cover the top with a thick cloth. Then they waited for the eggs to hatch. First one egg broke open and a hatchling stuck out its nose. Hatchling turtles have a little projection on the top of their snout that we call the egg tooth. It is sharp enough to rip through the leathery eggshell. This first hatchling just laid there and occasionally squirmed. Then another hatchling did the same and then another. Over a few hours, more and more hatchlings emerged. The first ones climbed out of their shells and stretched out straight. They had been rolled up in a sort of ball shape for 2 months and must have been a bit stiff. As the early hatchlings started to wiggle and crawl around, they bumped the unhatched eggs, and this stimulation got those hatchlings to wiggle, too. They broke out of their shells and more and more hatchlings became active.

Now the hatchlings found themselves in a chamber full of siblings with nowhere to go. As they crawled around and thrashed against one another, the ones on top bumped against the roof of the chamber and knocked down some sand, which trickled to the bottom of the pile. The hatchlings on the bottom smashed down the sand beneath them so that the sand from the ceiling now made the floor rise up. The pile seldom stopped moving, because when things got quiet those that were buried by their peers would start to wiggle around to get out from under the pile, thus reinvigorating the entire group. On it went, the hatchlings rising up together as if they were in a slow elevator.

This process goes on in spurts and stops, so it takes awhile to complete. The point when the hatchlings come out of their shells and start to get really active is when we measure the lowest amount of O_2 in the nest. It can drop to 12 or 13 percent, and that will shut off activity because the hatchlings cannot continue to fuel their metabolism with such thin air. So they stop and rest.

After a while the O_2 level goes up in the nest and one of the hatchlings will suddenly burst into action and squirm around as if trying to get out from under the pile. This activates the entire group, and they are off and moving again. So the entire process goes, with no particular plan or organization. It is an unorganized collaboration of individuals all doing their own thing but being pushed on by their neighbors. Not unlike the action of a human mob, this frenzied group of turtles is carried to the surface as the former egg chamber rises upward toward the open air.

Avoiding the Heat

One problem that the hatchlings face as they approach the surface is that it gets hot during the day and they can overheat. Thus they come out at night. Some nature films depict sea turtle hatchlings emerging from the nest in the middle of the day and, of course, they are met with hordes of birds that chow down on them as they run to the water. Most of those episodes are faked, set up for the cameras. I was involved in one such filming. Early in the morning, just as the sun came up, we buried some hatchlings a couple of inches in the sand and the film people got great footage of them as they popped out and ran to the ocean. In our case, the sand was cool and we guarded the hatchlings as they made it to the water and swam off. But that was for Hollywood.

I have only seen hatchlings naturally emerge in the daytime once, and it wasn't pretty. My students Laurie Cotroneo, Shaya Honarvar, Steven Pearson, and two Costa Rican students, Grettel A. Murillo and Luis G. Fonseca, who were working with Shaya on her olive ridley project, had hiked over from Playa Naranjo to Playa Nancite in Santa Rosa Park along the Pacific coast. Shaya was from the Netherlands and trained in molecular biology. She used that training to investigate the effect of microbes in the beach on the hatching success of olive ridleys. Nancite was an *arribada* beach, and Shaya was there to collect more samples of sand for her project. Laurie was a 3.75 student as an undergraduate focused on molecular biology. However, she did a study-abroad trip to Botswana, where she fell in love with field biology.

We were all walking down the beach and noticed six or seven frigate birds farther along diving to the sand. It was about four o'clock in the afternoon, and the button bush mangrove trees behind the beach were throwing some shade under them, so the sand was cooling off. In that shade some olive ridley hatchlings were emerging from their nests. They were in a hurry to get to the ocean. Unfortunately for them, they got about a meter toward the water from the shade and reached sand that was in the full sun at about 55°C (131°F). They took a few strokes with

their flippers and went stiff because they overheated. That was the banquet that the frigate birds were enjoying—broiled olive ridley hatchlings.

The students were concerned for the turtles. I delivered my lecture about natural selection and how the process was all for the best and we should leave it alone. After all, frigate birds need to eat, too. But the students were relentless, arguing for conservation, and I relented. So we started to gather up all the hatchlings that we saw. Several nests were emerging, and we had a bundle of hatchlings.

We had not brought anything to put hatchlings in because to get to Playa Nancite you have to make a tough climb over a small mountain and the students had plenty of scientific gear to haul. So we looked around on the beach and picked up all the containers that we could find. In this case, all of that junk that washed up on the beach actually came in handy. We found a Styrofoam cooler, some empty bowls, a wooden box, and a couple of other useful items that we filled with hatchlings. After dark we released the hatchlings, and they made a safe run to the sea. We wanted the turtles to walk down the beach themselves, because there is data suggesting that hatchlings smell the sand and remember the beach, returning to it years later to nest.

There has been a lot of speculation about the mechanism that causes hatchlings to emerge primarily at night. John Hendrickson suggested back in 1958 that high temperatures inhibited the emergence of green turtle hatchlings, but recent authors disagree. Graeme Hays has suggested that the rate of change in the sand temperature provides the cue for emergence. Kate Moran, a student of Karen Bjorndal at the University of Florida, took some measurements from which she concluded that there was a temperature threshold for emergence. One of my students, Dana Drake, recently did some key experiments to determine the effect of temperature on emergence of hatchlings. Dana measured the temperature tolerance of hatchlings by heating them up in water in a 2-liter flask and noting their behavioral responses. At 33.4°C (99.1°F) green turtle hatchlings became uncoordinated. Leatherback hatchlings did the same at 33.6°C (92.3°F), while olive ridleys tolerated a temperature of 35.7°C (96.3°F). Hatchlings generally went into spasms at 40–41°C (104–106°F), the temperature that we termed their critical thermal maximum (CTM). The hatchlings were not permanently harmed, because as soon as they reached their CTM, Dana put them into colder water and they quickly recovered.

Next Dana measured the temperature of wild hatchlings when they were close to the surface but not moving—34.4°C (93.9°F). When she found some leatherback hatchlings out on the sand at midday, their temperature was 36.5°C (97.7°F). The hatchlings crawled around for a few minutes and went into spasms when their body temperatures reached 41–43°C (106–109°F). She plunged them into cooler seawater at 29°C (82°F) and they recovered. Finally, Dana measured the temperature of hatchlings as they emerged from the nest at night. Those hatchlings that came out just after sundown were about 31°C (88°F), and those emerging later at night had a lower body temperature of 29°C (84°F).

Hatchlings in the field stopped moving at temperatures similar to those at which they became uncoordinated in the heating experiments. And hatchlings that emerged always had temperatures below 34–36°C (93–97°F) with sand temperatures below 36°C (97°F). Dana also found that the *rate* of change in sand temperature was too low for the turtles to use it as a cue. So threshold sand temperature was both the cue and the factor controlling when hatchlings emerged.

What we think happens to hatchlings as they crawl up from the nest chamber is, first, it appears that there is no particular time of day when hatchlings begin to hatch and when they start to crawl up from the nest. We can thus assume that that process can start at any time of the day or night. It is also reasonable to assume that hatchlings arrive near the surface throughout the day and night. If hatchlings arrive at night and the sand is cool, they come right out and they will have temperatures similar to those of the sand just below the surface. If they come near the surface during the day, they hit hot sand and become quiescent—they get stiff! Then as the sand cools off and the body temperature of the hatchlings drops below 34°C (93°F), they become active again.

Fortunately for the hatchlings, the ones on top cool off first since the wave of cooling in the sand starts from the surface and goes down. Otherwise the ones down below would get active first and push the top layer of hatchlings up into the hot sun. Therefore, the greatest numbers of hatchlings emerge in the first couple of hours after sunset because there is a backup of hatchlings near the surface waiting for the sand to cool off. During the remainder of the night, hatchlings can come out any time as soon as they come near the surface, which will remain cool until the sun warms up the beach the next morning.

The Crawl to the Sea: Dogs and Raccoons

As soon as hatchlings come close to the surface, they are exposed to a whole series of predators that are looking for an easy meal. Dogs, both owned and feral, are very efficient at finding hatchlings in nests before the turtles emerge. Dogs can smell the hatchlings from 30 m (32 yards) or even more. If there are dogs on the beach, sea turtle hatchlings are often evening and morning snacks.

As part of an extensive study in the early 1990s, Alison Leslie reported that dogs dug up a third of all leatherback nests at

Tortuguero before the hatchlings emerged. Alison graduated from Stellenbosch University in South Africa and after graduation worked her way around the world, picking kiwi fruit in New Zealand, commercial ice fishing for walleyes and trout in northern Canada, and looking for other interesting things to do. Alison is from English stock, majored in botany and zoology, was a star at sports including field hockey, spoke Afrikaans and Zulu, and had contagious enthusiasm. She applied as a volunteer for one of my projects, and we were somewhat apprehensive at having a white South African working on the beach. Almost the entire village of Tortuguero was of Jamaican heritage, and Costa Rica was boycotting South Africa. We decided to judge Alison by her abilities, not her nationality or race, so we had her come on down from the colds of Canada.

No one has ever worked harder or smarter on the beach. She did such a wonderful job as a volunteer that we invited her to do a master's degree with us. It was Alison's research that determined how much of a conservation threat dogs were to hatching sea turtles. Alison later completed her doctorate with me at Drexel University on the biology of Nile crocodiles in South Africa. She finally became a faculty member at her alma mater, Stellenbosch University, where she has served as department head and directs her own students studying crocodiles and sea turtles in several African countries.

Dogs are a serious problem on any beach where they are left to run free. At Playa Grande, Bibi Santidrián found that the rate of nest predation by dogs increased from 8 percent in 2005–2006 to more than 50 percent of nests in the years 2007–2009. And that beach is in a national park where turtles are supposed to have the highest protection. Imagine what it is like on other beaches in Costa Rica where protection is less formal.

In his classic book, *So Excellent a Fishe,* Archie Carr recounted the impact of dogs on the nesting of green turtles at Tortuguero. Packs of dogs sometimes came over from the mainland towns and ran wild on the beach during turtle nesting season. They dug up nests, ate hatchlings, and even caught eggs as the turtles were dropping them into the nests. Finally, park rangers shot all of the dogs they could find, including dogs that wandered out from the village of Tortuguero. That was tough medicine and upset many people, but it saved the nesting population of green turtles. Yet in 1990 and 1991 Alison found that the dogs had returned and were once again menacing the turtles. In this case it was leatherback eggs and hatchlings that the dogs ate, because the beach protection did not start on the beach until (later) green turtle season began. Due to Alison's discovery, nest protection now starts earlier and the leatherback nests are much better guarded.

There is a real problem in the United States, but rather than dogs, the issue is raccoons. Archie Carr predicted it in 1967. He wrote then that the biggest threats to the loggerhead population in the southeastern United States were raccoons and real estate development. As people built hotels, condominiums, and houses on and near the beaches, nesting areas were lost and lights disrupted adult and hatchling turtles. The raccoon population exploded because raccoons do very well living near people. They get supplemental food from trash cans and no longer have to fear their predators: pumas (aka mountain lions or cougars) and human hunters. Several studies have suggested that raccoons are more abundant now than in the past and on some coastal islands dig up most loggerhead nests. They are a plague and part of the reason loggerheads are declining in the United States. In many areas we need to make a choice: Do we want sea turtles or raccoons? Is killing raccoons defensible in an effort to save turtles? I would argue yes, we should drastically reduce the number of raccoons in areas where sea turtles nest, because raccoons are plentiful in many places and not in danger of extinction.

Birds, Coatis, Crabs, and Crocodiles

Sea turtle hatchlings rest on top of the nest for about 20 minutes after emerging, probably to let their muscles rejuvenate. They then take about 15 to 20 minutes to crawl from the nest to the water, moving at a rate of about 3 meters (about 10 feet) per minute. During that time they are beset by a diverse array of predators. Bibi Santidrián once used a military night-vision scope with a photomultiplier tube to watch the process of emergence and the crawl to the sea from start to finish in forty-seven nests. Bibi observed how long it took the hatchlings to move out of the nest, how fast they crawled, how straight they crawled, what predators were around, and how effective the predators were. Predators, she reported, caught 12 percent of leatherback hatchlings at night, and another 5 percent of hatchlings got stuck or flipped over and were probably eaten as well. So 83 percent of hatchlings made it to the water in this relatively dog-free zone, which is a much higher percentage than I would have guessed.

Bibi's study paints a picture of how things happen after the hatchling leaves the nest. The biggest problem is the smallest predator—the ghost crab. These sand-colored animals account for almost half of the predation. They pop in and out of holes in the sand. The crabs grab hatchlings by their flippers and drag them down into their holes, holding the hatchlings until they become exhausted from struggling to escape. Sometimes the hatchlings drown when the tides come in. Yellow-crowned night herons and great blue herons were also significant predators in Bibi's study, accounting for 25 percent of the predation.

As dawn approached, Bibi saw crested caracaras (indigenous

eagles) patrolling the beach and picking up hatchlings that emerged near sunrise. Frigate birds came by early after sunrise and ate hatchlings that were swimming in the water near the beach. Pelicans took a few as well.

There were no raccoons or coatis on the beach during Bibi's study. A few years earlier we often saw raccoons on the beach and some coatis as well. In the early 1990s there were coatis but no raccoons. Then one raccoon mother began to go to the beach and dig up nests and eat hatchlings. After a couple of years there were more raccoons, as her young learned from their mother. Even 3 or 4 years before Bibi's study there was a sizeable raccoon population working the beach, and some even crawled over the fence to try to get into our hatchery.

About that time more construction began around Playa Grande and the raccoons disappeared. Big lizards, such as spiny-tailed iguanas, and some of the monkeys also disappeared. The local people said that the workers, who were mostly illegal immigrants from Nicaragua, were eating all of the animals. We had no way to verify this claim, but the animals did start to disappear when the workers arrived. It is common for Ticos to point fingers at Nicas (Nicaraguans), so it is hard to tell truth from myth. Coatis were still around the area and some raccoons as well, but they did not come onto the beach. Maybe all of the tour groups, rangers, and biologists discouraged their activity, or maybe picking garbage was easier.

In other parts of the world a wide variety of predators eat hatchlings on beaches. Ghost crabs certainly lead the list, as they are active worldwide. In the United States we have to add pigs to the dogs and raccoons. Coyotes are active on many beaches in the Americas, and jaguars are returning to Tortuguero, where they eat adult green turtles as well as hatchlings. In Australia, introduced red foxes create havoc on the beaches, but an eradication program is showing promise in reducing their numbers. Dingos are natural predators there, and rats are a scourge in Australia and on many Pacific Islands. Varanid lizards called sand goannas, relatives of the Komodo dragon, are very effective predators and take 50–67 percent of flatback turtle hatchlings at Fog Bay in Australia. A wide variety of birds make a living on sea turtle hatchlings as well. These include vultures—especially on olive ridley beaches—herons of several species, storks, pelicans, sea eagles, ospreys, kites, and other species, including crocodiles.

Lights: The Final Insult

As if all of these challenges facing hatchlings are not enough, humans have added another problem for hatchlings—lights. Many turtle nesting beaches in Costa Rica are still dark because they are in national parks like Tortuguero and Santa Rosa. At Parque Nacional Marino Las Baulas, where we study leatherbacks, the main beaches are relatively dark because of the vegetation barrier behind the beach and the limited number of houses nearby. On the north end of Playa Grande there are houses right on the beach and bright lights shine out to the water. This causes problems for hatchlings because they get disoriented and crawl to the light instead of the ocean.

Nearby Tamarindo has an even bigger problem because it is so bright the lights hurt your eyes when you look over there from Playa Grande at night. Hotel owners in Tamarindo hire small boys to walk the beach in the early morning to pick up dead and disoriented hatchlings before the tourists wake up to swim.

Archie Carr wrote in 1967 that lights were becoming a big threat to loggerhead nesting in the United States. Before the harnessing of electricity, the ocean was always brighter, due partly to the fact that the vegetation (including trees) on the land side of the beach tended to absorb light and block the sky. Now lights from hotels and houses, street lights, and cars on roads—all of which are on the land side of the beach—disorient hatchlings, and even adults. Hatchlings are drawn away from the water and onto the roads, where they are run over by the thousands. The problem has been addressed on some beaches in the United States but still continues to plague turtles in many places. There is a constant, but winnable, struggle to reduce the lighting.

A quality sea turtle nesting beach means that development must be kept under control and the area must be dark. The array of national and state-protected refuges in Florida are the places in which most successful nesting occurs. In other states like South Carolina, recent developments on islands such as Kiawah have taken the light pollution problem into account and the houses are behind the sand dunes. Further, lights are turned off at night. Kiawah is an example of the right way to develop. Keep the beaches dark and keep them quiet, control the dogs and raccoons, and then the hatchlings will have a fighting chance to reach the waves.

Rolling Waves

In addition to being programmed to head for the lightest part of the sky, hatchlings also sense the waves. Sound or vibration caused by the waves pounding ashore seems to provide an important signal. Many people have observed that hatchlings head for noise—the vibrations are actually carried through the sand. Nearby air conditioners, pool pumps, and diesel generators make sounds that are similar to waves, and they sometimes disorient hatchlings. I once followed a hatchling's track from its nest, off of the beach, through the coco plum and the tall grass, up onto a grass lawn, and to the edge of a swimming pool by a house. There I saw a hatchling vainly trying to reach the pool pump. I carried the baby turtle down to the edge of the sea to convince it that the sound of the waves was more enticing than the sound of the pool. No one has yet demonstrated in the laboratory that hatchlings respond to sound, so there is more research to do in this area. Maybe, for example, it is the vibration carried through the sand and not the sound itself that is important. Either way, data seem to be mounting that sound is important to hatchlings looking for the sea.

4

TO THE HORIZON: THE FIRST DAY

The adventure begins anew when the surviving hatchlings reach the ocean. When the first wave hits the tiny turtle, it is turned around a bit. It also takes its first drink of water, gulping it down. Then the hatchling rights itself and heads on into the breakers. After a wave or two the hatchling finds itself afloat in the sea and it begins to swim, moving its front flippers up and down together the same way that a bird moves its wings in flight. The hatchling is literally flying through the water, coming up to the surface to breathe and diving to a meter or more in depth as it is carried out to sea by the retreating undertow. It is able to maintain its seaward direction by orienting into the waves until it is well beyond the surf zone.

As it swims the hatchling keeps drinking to regain the lost 20 percent of its weight. A sea turtle excretes a salt solution in its tears that is more concentrated than seawater, so the hatchling can gain one drop of pure water for every two drops of salt water that it drinks. Humans don't have a salt gland and thus cannot drink seawater. For hatchlings the old line might be changed to, "Water, water everywhere and plenty of drops to drink."

Hatchling Migration

As a hatchling swims into the waves, the shore begins to fade. This is a remarkable feat that surpasses the migration achievements of birds. When birds undertake their first migration, they are months old and have already been practicing their orientation skills with short flights around their nest area. The birds have had those months to learn visual cues in their environment such as sun location, day length, and major landmarks. In an hour or so a turtle hatchling emerges from underground at night, finds the ocean, and sets off.

Over the last 50 years, several scientists have contributed to our understanding of how hatchlings manage this feat. Archie Carr and his student Larry Ogren were among the first to conduct an experiment on this subject. They transported green turtle hatchlings from the Caribbean coast of Costa Rica to the Pacific coast and let them go on the beach. They found that the hatchlings marched right into the water and set off west into the ocean, in the opposite direction from the one they would have taken at Tortuguero. Thus the hatchlings were not programmed to go in a specific direction; that would have sent them eastward.

Jane Frick was the first person to try following hatchlings in the ocean. She worked in Bermuda and was a strong swimmer, so she was able to follow hatchlings for 1–4 hours. She discovered that hatchlings maintained a steady course away from shore even after they moved over the horizon and could not see any landmarks. More recently, Blair Witherington, Mike Salmon, and Jeannette Wyneken did some very interesting studies of hatchling orientation on the beach and in the nearshore waters of Florida. These Florida biologists conducted both observations in nature and experiments in the laboratory to unravel the mysteries of turtle orientation. Blair tracked hatchlings

for hours and days as they swam off into the open ocean. Mike and Jeannette conducted some clever experiments in which they tethered hatchlings in a kiddie pool in the laboratory and observed their swimming behavior and orientation.

One of Mike's early collaborators, Ken Lohmann, made even greater strides in this arena. Ken, a tall, earnest-looking guy with a happy smile, learned the sea turtle business doing experiments in the field and in a wave tank in the laboratory with Mike and Jeannette. Together they discovered that hatchling loggerheads appeared to orient into the waves during their offshore migration. Then they tethered loggerhead, green, and leatherback hatchlings in a large tank in the lab and exposed them to simulated waves. When there were no waves the turtles oriented randomly, but when the waves were turned on the hatchlings swam into the waves.

Magnetic Personalities

Since the 1990s, Ken Lohmann and his wife, Cathy, have blazed new trails in unraveling the mysteries of sea turtle orientation. Cathy and Ken both earned their PhDs at the University of Washington, where Cathy studied the molecular response of insects to heat shock and Ken studied neurobiology, and now they work together as faculty members at the University of North Carolina. They have discovered that hatchling sea turtles can orient into waves even in the absence of light and that hatchlings respond most strongly to wave motions that closely resemble those of typical waves at their natal beach.

Some of their cleverest experiments have documented that loggerhead and leatherback hatchlings can orient to the earth's magnetic field. They were able to create magnetic fields in the lab by building a special magnetic coil system consisting of an array of wires. Then they tethered hatchling turtles in a swimming tank—a bit more sophisticated than the kiddie pool—and allowed the turtles to swim into the waves coming from a particular direction. They turned off the waves, and the turtles continued to swim in the same direction. When they reversed the magnetic field, using their coil system, the turtles turned around and swam the other way! Hatchlings, it seems, get their magnetic compasses set when they are crawling on the beach and swimming into the waves, and they keep that heading as they swim far out to sea. Under natural conditions, hatchlings use visual cues to find the ocean and orient into the waves and offshore. This course is then transferred to their magnetic compass so that the hatchlings can continue along the same heading after swimming beyond the waves that give them an early directional cue.

Tracking Little Turtles

Leatherbacks that hatch on Playa Grande and other Pacific beaches of Costa Rica swim out to sea in the same manner as loggerheads in Florida. They orient into the waves and head off to the southwest. In December, when the first hatchlings emerge, it is still rainy and stormy, so it is hard to do experiments. In January the rains have stopped, but the Papagayo winds appear and can blow at 70 km (50 miles) per hour or more for days at a time. These winds are the result of a pressure difference between the Caribbean Sea and the Pacific Ocean. As the winds stream down off the mountains, they roar out to the Pacific Ocean. When that happens you risk being blown far out to sea if you have boat problems. In addition, there are hidden rocks off the beach, and it is hard to navigate around them at night. For those reasons it took some years to find someone to follow hatchlings as they left Playa Grande.

Andy McCollum is a professor at Cornell College in Iowa—

the original Cornell, despite what New Yorkers may think. Andy had been carrying out a study of box turtle nesting and invited me out to give some lectures at the college. We discussed his interest in studying the behavior of sea turtles, and I described the need for a study of leatherback hatchlings. I "forgot" to tell him about the rainstorms and Papagayo winds. He jumped at the chance to spend his sabbatical doing those experiments at Playa Grande. Over two seasons he was able to track a number of leatherback hatchlings despite the misgivings of Christian, the local captain we hired to run the boat. Luckily, they did not run into any rocks.

Andy and Christian would set out from the nearby port of Flamingo in the afternoon and motor the few miles to Playa Grande. When it was dark, Andy's student Tera Dornfeld would release a hatchling on the beach and Andy and Christian would watch it by following the little light on a float the turtle towed behind it. The float was tied to a fishing line that attached with breakaway thread to a little Lycra harness on the hatchling. For the next few hours Andy and Christian would follow the hatchlings as they swam out to sea. Most of the hatchlings made it out a few miles, where they entered the north-flowing Costa Rica Coastal Current. Some of the night hatchlings became fish food. Once in the current the turtles drifted north.

Combining Bibi Santidrián's research that more than 80 percent of hatchlings that make it out of the nest get to the ocean, and Andy McCollum's work that 50 to 75 percent of those hatchlings make it out to the Costa Rica Coastal Current, it looks like about half of the turtles live to see their first sunset.

5

LOST AND FOUND: LIFE AS A JUVENILE

The land is gone. The hatchling is beyond the horizon, surrounded by water, water everywhere. It was not eaten as an embryo. It did not get flooded or washed out to sea while still inside the egg. It made it from the nest to the water. It made it past the first set of predators. This tiny creature, 5 cm long (2 inches), is on its own but still alive.

If the turtle is a loggerhead that hatched in Florida, Georgia, or the Carolinas, it will not be seen again in the coastal waters of North America until its shell is 50 cm (20 inches) long. Archie Carr famously referred to the time between seeing hatchlings leave the beach and noting their return as older juveniles as "the lost years." For decades no one knew where the turtles went when they left the beach. At first Archie thought sea turtles spent a single "lost year" somewhere out in the ocean. The prevailing idea was that it took only a year for a hatchling to grow large enough (dinner plate or platter size) to make its way back to coastal lagoons and estuaries. We now know it takes much longer.

In his 1967 book *So Excellent a Fishe,* Archie described how he set about looking for these "lost" juveniles. First, he searched the shores of the Caribbean, the Gulf of Mexico, and the U.S. Atlantic coast for any signs of very little sea turtles. He did not find any. Then he systematically canvassed fishermen who set or dragged nets along the shore, in bays, and near the beaches during and after the turtle hatching season. Again they did not catch any small turtles. So Archie concluded that the turtles must leave the beach and head out to sea to find food and avoid the many fish predators that live in the nearshore environments. He noted that the coloration of many species of sea turtles, dark above and white below, provided the same kind of countershading typical of free-swimming open ocean fish. The white underbelly makes the fish or turtle less visible against the sky to a predator viewing it from below, and the dark back makes it less visible to birds overhead, as it merges with the darkened color of the ocean.

The First Hints

Over the last 40 years numerous scientists have, through long and arduous work, clarified where sea turtles spend their juvenile years and what they do there. One of the first clues came from reports by Leo Brongersma that small loggerheads, with a carapace (or top shell) of up to 40 cm (16 inches) long, washed ashore in Europe. Brongersma, a Dutch herpetologist who became director of the National Museum of Natural History (Rijksmuseum van Natuurlijke Historie) in Leiden, Netherlands, was born west of Amsterdam and became interested in amphibians and reptiles while a student at the Municipal University of Amsterdam. He had a long career as a classical zoologist studying the taxonomy and anatomy of amphibians and reptiles in Europe and what is now Indonesia (at that time the Netherlands East Indies) and New Guinea.

Brongersma's contributions to the mystery of the lost years of sea turtles derived from his extensive studies of all of the available records of sea turtles beached (stranded) or sighted in

Europe. By examining popular as well as scientific articles, he became convinced that small sea turtles were an important part of the European fauna and wondered where they came from, how they got there, and whether arriving in Europe meant they were hopelessly lost. He also became interested in sea turtle conservation but was convinced that sea turtles were a valuable resource to farm and exploit. This led to serious conflict with many sea turtle biologists, including Archie Carr. Archie had started out thinking that sea turtle farms could be a useful conservation tool but became convinced that they were in fact a negative factor because they relied on taking animals from the wild and were a source of increased demand for sea turtle products. Most people now think Carr was correct, but this discussion has continued long after Archie's death in 1987, as evidenced by current debates about the Cayman Turtle Farm and other such operations. Brongersma's long and important career is more often remembered among sea turtle biologists for this conflict than for his many scientific contributions, including the fact that his summaries of the turtle records in Europe helped to solve the "lost year" puzzle.

Driftlines and Convergence Zones

Wherever two currents meet in the ocean, water collides and sinks. Any material that floats, such as floating algae, is drawn together in a line. These driftlines, or rips, are the ocean's grocery stores. The accumulation of plankton, copepods (small crustaceans), salps (gelatinous animals), and other jelly creatures is mixed with old trees and logs on which barnacles and clams make their living. Small fish hide under the logs, and bigger fish seek them out for a meal. If you take a small plane from Miami to Bimini in the Bahamas, you can see the weedline along the front of the Gulf Stream for miles, and the birds

circling overhead give it away to an observer on board a ship. In the eastern Pacific Ocean from Cocos Island to the Galápagos Islands, boobies, frigate birds, gulls, petrels, and albatrosses swoop and dive into the rich soup of prey along fronts that stretch for hundreds of miles as the North Equatorial Counter Current flows east past the west-flowing South Equatorial Current.

As the warm Gulf Stream wanders north of Cape Hatteras, North Carolina, it encounters cold water coming south, and the mixing action causes rings of warm and cold water to break off and spin away. Warm-water rings bring tropical fishes to New Jersey and New England, and cold-water rings bring northern fish to the area south of the Gulf Stream. These rings can stretch for miles or be as small as a few hundred meters in diameter. While the big convergences are the grocery stores or even Wal-Marts of the ocean, the small rings are sort of like the 7-Eleven or WaWa convenience stores of the ocean, where a bird can drop in for a quick snack.

In 1986 Archie Carr published a very important scientific article in the journal *BioScience* called "Rips, FADS, and Little Loggerheads." In it he highlighted the importance of downwellings of seawater in producing fronts or lines along the surface of the ocean where *Sargassum* and other resources converge in long streams, or driftlines, that stretch for miles. (*Sargassum* is a genus of large brown seaweed—a type of algae—that floats in island-like masses and is common in waters off the U.S. South Atlantic coast.) Here nutrients accumulate, plankton collect, small fish show up, and birds and dolphins congregate. Food is actually scarce in the big ocean, and these driftlines are like oases. In the open ocean it might take a tiny turtle days or weeks to find enough food to fill its stomach. Along a driftline it might take hours. And there, it turns out, is where the lost turtles go.

The Azores Connection

In the Azores Islands, two-thirds of the way across the Atlantic Ocean off Portugal, fishermen on tuna boats used to catch little loggerheads with hand nets and take them home to eat. Decades ago a young marine ecologist named Helen Rost Martins contacted Archie Carr about the situation. They began a collaboration in which she bought the turtles from the fishermen, measured them, attached flipper tags, and set them free. When the turtles they marked were recaptured later, an amazing fact was learned. The lost year, it turned out, was more like a lost decade.

Helen and Archie's efforts continue to this day, but with the help of Alan Bolten and his wife, Karen Bjorndal. Alan grew up in New Jersey, a state where teenagers identify more with its turnpike and parkway exits than with the extraordinary wonder of its Pine Barrens and estuaries. Alan was the exception; following his interest in nature to the University of Florida, he became a leading sea turtle scientist. Karen grew up in the San Francisco Bay area and has been enthralled with turtles since she received a "dime-store," red-eared slider from her aunt Lillian. She is unquestionably the world's leading expert on green turtles and a leader in the sea turtle conservation community. Always energized, Alan and Karen were key in starting the Archie Carr Center for Sea Turtle Research in the university's Department of Biology.

The Martins-Bolten-Bjorndal collaboration has been extremely fruitful. Helen measured the first forty loggerheads that were brought in by the fishermen and found that they filled in exactly the gap in size (20–40 cm / 8–16 inches) between small turtles found in *Sargassum* weedlines off Florida (5–20 cm / 2–8 inches) and the bigger juvenile and subadult turtles (40–90 cm / 16–35 inches) that were found in the lagoons along the East Coast of the United States. Alan worked with Martins in the Azores to increase the number of turtles that could be measured and tagged. Like Helen, Alan worked with the local fishermen.

Most tuna fishermen in the Azores are traditional hook-and-liners and rely on "fishing the birds." They watch for groups of birds feeding at the surface along convergence lines that indicate where tuna are chasing small fish near the surface. Sometimes they see turtles floating on the surface and can motor up to them and catch them with a net. Working with the scientists, the fishermen attach metal flipper tags to the front flippers of the turtles and release them. Specially trained fishermen also measure the turtles. Some cooperating fishermen bring turtles to Martins so she can make more extensive measurements before tagging and releasing them.

Recaptures of these turtles have allowed Martins, Bolten, and Bjorndal to calculate growth rates and determine the length of the lost year period. The turtles spend about 6–12 years at sea.

Finding Nemo: How Juvenile Turtles Find Their Way

In 2003 a wonderful animated children's movie captivated audiences as Marlin, a small tropical clownfish, undertook a great migration to rescue his son, Nemo, from a fish tank in Sydney, Australia. In *Finding Nemo,* the main character is helped along the way by a group of loggerhead turtle "dudes" that are surfing a current on their migration through the ocean. Do juvenile turtles actually surf their way around the ocean? Well, in a way, yes they do.

Off the Florida Atlantic, Georgia, and Carolina coasts most hatchling loggerheads end up in the Gulf Stream. There they find a rich soup of food that includes small salps, ctenophores ("comb jellies"), and zooplankton, as well as less nutritious debris such as bits of plastic and oil droplets. Eventually they make their way to mats of floating *Sargassum* algae that are home to a hundred or more species of plants and animals. These mats are almost like islands that float on top of the ocean. Young loggerheads eat a variety of things that can be found on *Sargassum*—small barnacles, tiny crab zoea (larvae), fish eggs, and hydrozoan colonies. They also eat things that have blown from land—ants, aphids, leafhoppers, and beetles. The little loggerheads may develop a geomagnetic map of the ocean as they drift. They live on or near the surface and hide in the *Sargassum* to avoid predators. These little loggerheads travel the North Atlantic Gyre, which transports them across the North Atlantic toward Europe. Some of them drift south past the Azores. Others, as they approach Portugal, meet a northern branch of current—one that drives northward past the west coast of Great Britain. The southern branch sweeps toward the South Atlantic. For this year's crop of hatchling loggerheads, who have taken a month or more to reach that point, a wrong move is fatal. Those loggerheads that go north soon die from the cold. Those that go south get swept away from food. To survive, they must stay in the current and circle the Sargasso Sea in the center of the North Atlantic Ocean.

Ken and Cathy Lohmann made an important contribution to sea turtle biology by figuring out how those little loggerheads manage that feat. In another series of experiments on hatchlings' magnetic responses, they discovered that loggerheads respond to different inclination angles in the earth's magnetic field. Those field-line inclinations differ at different latitudes. The Lohmanns used a computerized coil system, generated with vertical and horizontal power supplies and coils, to simulate different inclinations in an earth-strength field in a water-filled arena. Again, it was electrical engineering magic. Hatchling loggerheads exposed to an inclination angle like that at the

northern edge of the gyre swam south-southwest, and those exposed to an inclination angle like that at the southern edge of the gyre swam in a northeasterly direction. These behaviors tend to keep loggerheads safely along the edge of the Sargasso Sea.

Then Ken and Cathy changed the intensity of the magnetic field. When hatchlings were exposed to a field intensity like those off the eastern United States, they swam in a northeasterly direction. This would keep them in the Gulf Stream or take them out of it in the direction of the North Atlantic Gyre. When hatchlings were exposed to a field intensity like that found off Portugal, where the gyre splits, they swam west, which would take them away from Great Britain and away from the southern sweeping current and keep them in the gyre.

Finally, they exposed hatchlings to magnetic fields actually found in three widely separated locations along their migratory route in the North Atlantic Gyre. Turtles exposed to a field replicating one that exists offshore near northern Florida swam east-southeast. Those exposed to a field like that near the northeastern edge of the gyre swam south. Those exposed to a field like that near the southern part of the gyre swam west-northwest. So hatchling loggerheads have a guidance system in which regional magnetic fields act as navigational beacons and the hatchlings change swimming direction as they come to different geographic boundaries. Through years of natural selection those turtles that did not have that ability or had it to a lesser degree got swept away from the gyre and ended up in cold water to the north or warm water to the south that did not contain enough food. They died and did not pass on their genes. Those that had the ability lived to pass on their genes. So the survivors have a remarkable ability to navigate across the Atlantic Ocean.

Turtles in other parts of the world have much different developmental habitats. They presumably use the same magnetic abilities to find their way to a safe place to grow up, but they face different oceanic conditions, ranging from currents to climate. One of the main nesting areas for loggerheads in the Pacific Ocean is Japan. Hatchling loggerheads that leave the Japanese beaches get swept into a current system that takes them north and around the Pacific. The water is much colder, and that journey presents some additional challenges to those little turtles because it is longer than the trip in the Atlantic. Eventually the turtles end up along the west coast of Mexico, where they spend their juvenile years.

The oceanic zone is the vast expanse of open ocean where the depth is greater than 200 m (650 feet). Here juvenile sea turtles actually do ride, or surf, the Gulf Stream and other currents as they drift around the North Atlantic. It would take about 240 days for a hatchling to surf the current of the Gulf Stream from Cape Canaveral in Florida to the Azores.

Life in the Middle of the Ocean

It was a real surprise to most biologists when it became clear that juvenile sea turtles spent so many of their juvenile years out in the open ocean. We just never thought that those animals would be roaming the open sea feeding on little animals that floated along the surface layers of the ocean or that were attached to floating objects in those driftlines. Sea turtle biologists tended to think of turtles as living along the shore and did not conceive of them really as ocean creatures, even though we called them sea turtles!

Bolten and Bjorndal and their colleagues found that juvenile loggerheads out in the Atlantic spend 75 percent of their time in the top 5 m (16 feet) of the ocean and only occasionally dive to as deep as 200 m (650 feet). In areas where there are shallow underwater seamounts, ocean banks, and ridges, oceanic loggerheads may dive to the bottom to feed. Once little loggerheads reach the area around the Azores, they stay there until they begin their migration back to the North American mainland. On the other hand, loggerheads that swim past the Azores to Madeira off the northwest coast of Africa do not tend to remain there as residents. When released with satellite transmitters attached, several juvenile loggerheads swam north and west, which would take them toward the Azores.

For other species of turtles and for loggerheads in other places, we have scraps of information. For example, loggerheads in the Pacific Ocean hatch out from beaches in Australia and Japan. Yet small Pacific loggerheads, which grow more slowly than Atlantic loggerheads, appear to spend their oceanic life stages off North America, especially Baja California, and along the ocean fronts between currents in the middle of the North Pacific. Some live near Australia, Hawaii, and Peru.

The Pacific is a big ocean, and there is a need for many more studies to understand what loggerheads are doing out there during their oceanic stage. Unfortunately, the numbers of loggerheads nesting in the western Pacific are declining, and that indicates that there may be some serious problems for little loggerheads there. What is affecting them? Is it fishing? Hopefully, we will discover the problems and solutions before we run out of turtles to study.

Kemp's ridleys appear to have a similar lifestyle in the open ocean as loggerheads, but for a much shorter time span. Some hatchlings ride the Mexican Current off their main nesting beach at Rancho Nuevo, Mexico, and probably spend their oceanic phase in the large eddy systems in the Gulf of Mexico living in mats of vegetation and eating the animals associated with that ecosystem. Others pass out of the Gulf through the

Straits of Florida and into the Gulf Stream. Kemp's ridleys are seldom seen at sea and do not reappear along the coast until they are a few to several years old and reach a carapace length of 20–25 cm (8–10 inches).

Olive ridleys disappear when hatchlings leave the nesting beaches. Their early years are probably spent in the open ocean around driftlines, since that is where larger individuals are sighted, swimming around floating logs, trees, seaweed, and debris. Little olive ridleys eat the animals associated with these structures—snails, clams, and crabs. Atlantic Ocean hawksbills spend their lost years at sea in *Sargassum* and come back to the nearshore environment when they are 20–25 cm (8–10 inches) in carapace length. In the Pacific they move to coral reefs from the open ocean at a larger size, 35 cm (14 inches) or more, and at an older age. Flatbacks in Australia do not appear to have any oceanic life stage. They stay in the coastal zone and feed in the upper part of the water column.

Green turtles spend their early years at sea in the oceanic zone. They are found in convergence zones like other sea turtles but are only occasionally associated with floating objects. Small green turtles are found in *Sargassum,* but larger juveniles are seldom seen there. Yearling green turtles have been found in the area of Madeira, and they may have come from the west coast of Africa. In the open ocean, green turtles are omnivorous, eating snails, small ctenophores, and other animals. They enter the nearshore environment and switch to eating plants when they reach a carapace length of 20–35 cm (8–14 inches), depending on where they are from.

Leatherbacks have evolved to live completely in the open ocean. Their hatchlings are bigger than those of other sea turtle species and have proportionately longer flippers. They are true swimming machines. Once they swim away from the nesting beach, hatchlings are seldom seen again, and we have very little information about how they spend their early years. They are not seen near shore, so we assume they must be in the oceanic zone.

Scott Eckert, a research professor at Duke University and an expert on biotelemetry, has compiled all of the records of juvenile leatherbacks found around the world. Of the ninety-eight recorded animals he tallied, only seven were smaller than 20 cm (8 inches) in carapace length, and two more were between 20 and 40 cm (8–16 inches). All but one of these animals were seen in the tropics; the exception was at about 30° latitude. All were in the northern hemisphere and all were in water that was 26°C (78.8°F) or warmer. Large leatherbacks can thermoregulate, or control their body temperature, by using their large body size, thick, fatty insulation, and control of blood flow to their skin. Juvenile leatherbacks are not big enough to do that and seem to be restricted to warm water.

Encountering a young leatherback is a precious moment. One extraordinary video exists of a juvenile leatherback caught by a Costa Rican longline fishing boat. The turtle is about 50 cm long (20 inches) and so fat that it looks like a black and white football. It was very well fed. A Costa Rican biologist was on board as an observer to record the interactions of the fishermen and sea turtles as part of a program organized by Randall Arauz. Randall is the leading sea turtle biologist and activist in Costa Rica and is head of a very effective advocacy group called PRETOMA. His biologist shot the video as the turtle was brought on board, after which the crew removed the hook and set the turtle free. The boat was fishing between the mainland and Cocos Island, a hot spot for the migration of adult leatherbacks away from the nesting beaches of Guanacaste. It may be a great spot for juvenile leatherbacks as well. The waters of the Costa Rica Coastal Current sweep north past the mainland and swing back to the west. The Costa Rica Dome sits between these currents, and juvenile leatherbacks utilize the convergence zones along it to feed on salps, ctenophores, and small jellyfish.

Coming Inshore

As they get larger and larger, loggerheads in the Azores grow at slower and slower rates. That means that they cannot find enough food to maintain their early increases of 4–5 cm/year (1.6–2 inches). Eventually they will slow down to 1–2 cm/year (0.4–0.8 inches). On the other hand, loggerheads that migrate inshore to take up residence along the U.S. East Coast and other habitats around the world grow twice as fast. By the time the turtles reach 45–65 cm (18–26 inches) in carapace length, almost all of them have left the open ocean and taken up residence along the coast.

As they grow, green turtles shift from eating high-calorie animal food to less nutritious and harder-to-digest plants along the shallow waters. In the Caribbean they eat primarily sea grass, especially *Thalassia testudinum* ("turtle grass"), and in other places they might eat algae instead.

Hawksbills eat sponges on coral reefs in the tropics. They shift from eating an omnivorous diet along driftlines at sea, to an omnivorous diet when they first reach coral reefs, and then to a more specialized diet of all or mostly sponges.

Juvenile Kemp's ridleys feed along the East Coast of the United States as far north as Massachusetts, as well as along the Gulf coast into Mexico. Their close relatives, the olive ridleys, seem to be of two minds. While some juveniles move inshore and take up residence in the neritic zone, others stay out at sea and exploit the driftlines, continuing to feed on crabs, snails, barnacles, and jellyfish. They might even catch a small fish or two.

The other two sea turtle species never shift their habitats.

The flatback never leaves the nearshore zone along the continental shelf of Australia, so it does not migrate. The leatherback never gives up its diet of soft, gelatinous prey such as salps and jellyfish. It continues to hunt them around the world, eventually ranging as far north as Labrador and Alaska. Leatherbacks go where the jellyfish are, whether that is Monterey Bay, California, or to the depths of the ocean off the U.S. Virgin Islands.

Live, from New York

Most people are unaware that some of the real sea turtle hot spots in the world are right along the shore of the eastern United States, even in the waters around New York City. Ed Standora and Stephen Morreale, along with Ed's student Vincent Burke, carried out some fascinating studies of the rare juvenile Kemp's ridleys in Long Island Sound and other waterways around the Big Apple. They tracked the turtles with radio transmitters and found that they spend their time in shallow water, even swimming into marinas on occasion. Kemp's ridleys ate spider crabs, rock crabs, lady crabs, mollusks, and even sea horses. As winter approached, most turtles headed south and avoided cold water. A small percentage of turtles stayed too long and got caught in chilly waters. Too cold to swim, they were cold-stunned, a condition that causes them to float and become immobilized (often deadly). On the other hand, turtles that got out in time actually grew faster than their southern counterparts.

In Chesapeake Bay there are between 5,000 and 10,000 loggerheads each summer, most of them large juveniles. Jack Musick, a professor at the Virginia Institute of Marine Science, and his students have documented the biology of sea turtles in the Chesapeake by counting them from airplanes, getting them from pound nets set by fishermen, and tracking them with radio and satellite transmitters. Much of this research was done by Jack's student, John Keinath. Loggerheads enter the bay in May or June, when the water warms up to 16–18°C (60.8–64.4°F), and spend the summer there feeding along the bottom, especially along the edges of channels. In the fall, the turtles head south and are past Cape Hatteras by December or January. They return the next year and often come back to the same area to feed.

I have done some work in this area, particularly with loggerhead turtles in Delaware Bay. In 1995 John Keinath and I calculated that loggerheads were present in Delaware Bay at the same density as in the Chesapeake. We counted them from a small, low-flying plane (150 m/490 ft) and set nets for them from a boat. There were so many loggerheads in Delaware Bay that we even observed them, from time to time, stealing the bait from people fishing on party boats. Occasionally a loggerhead would get caught, a nasty surprise for anyone trying to unhook them, given their demeanor. Loggerheads and Kemp's ridleys will summer as far north as Cape Cod Bay. Rare records show these turtles even farther north, into Canadian waters.

The Leatherback, the Sailor, and the Nurse

Leatherbacks migrate up the U.S. East Coast from Florida starting in February. By late April they can be seen off the coast of Virginia, and they are observed off Massachusetts in August. Fishermen spot them off the coast of New York and New Jersey all summer long. They are often seen off Nova Scotia in summer, feeding on large schools of jellyfish. One of the most memorable moments I have had was seeing my first leatherback in Rhode Island back in 1980. It was a surreal experience. I was a professor at State University College at Buffalo and one day Ed Standora, then my research associate, came to my office and said that he got a call out of the blue from a guy named Chris

Luginbuhl. That started an adventure that has continued off and on to this day.

Chris said that he was head of the David Luginbuhl Foundation, named after his late father, and wanted us to come to Rhode Island to study leatherback turtles with him, Bob Shoop, a professor at the University of Rhode Island, and a young John Keinath, who was Bob's master's degree student at the time. They wanted to put transmitters on leatherbacks and follow them into the ocean. I looked up Chris's foundation and could not find any information. Even the Better Business Bureau had no information. I asked my brother John, an attorney, about it, and he told me to watch my wallet. So I was in a quandary—what to do?

Chris sent two round-trip airline tickets for Ed and me. We figured that at least we could get there and back, so we hopped on the plane and were met at the airport by Chris. Soon we were at a big, attractive house overlooking the ocean near Newport, Rhode Island. It belonged to a nice elderly lady who liked turtles, and Chris had convinced her to put us up for a couple of nights. That evening we met with Shoop and Keinath in a tavern and made our plans. Ed and I knew how to attach transmitters, and Chris, Bob, and John said they could catch a turtle.

At 4 a.m. we were on the dock having breakfast with fishermen at a coffee shop across the street from Macklin's Fish House. The fishing crew was going off to tend their pound nets, and they told us there was a good chance that they would catch a leatherback. We met them early to reassure them that we would neither get in the way nor were we crazed environmentalists who would disrupt their fishing business. It turned out that Chris was good friends with George Mendonca, the owner of the fishing business and, unknown to us, a celebrity. So far so good!

A few hours later George and his crew came back into port

with a boatload of fish: small tuna, mackerel, and other inshore fare plus what to us was a huge leatherback turtle in the middle of their deck. It was a large subadult, 103 cm (41 inches) long carapace length, and was healthy and eager to get back into the water. After the fishermen weighed and iced down their catch, they brought the leatherback onto the dock, placed it on a couple of tires so the rough wood would not scratch its plastron (or bottom shell), and then we pushed it into the fish house to keep it out of the sun. It was beautiful. It had a soft skin that you could scratch with your fingernail; it felt like a dolphin's skin but was actually more fragile. Black with small white spots, it glistened when wet. The shell featured five ridges running front to back. It was streamlined and was truly a magnificent animal. Later in life, when I began more intensive studies of leatherbacks, I wondered if this was a pivotal moment in my development as a scientist and conservationist.

We attached sonic and radio transmitters to the turtle, and the fishermen kindly offered to take us out to the edge of the harbor to release it back into the sea. John came alongside with his father's 45-foot Chris-Craft cabin cruiser, and we spent the next 24 hours tracking the turtle as it swam off toward Block Island. It was truly amazing. We could hear the transmitter and often knew that we were within 100 m of the turtle, but we could not see it. It was perfectly camouflaged, and the only way we found it was to move within 50 m (164 feet), when we could see the orange-colored float with the radio transmitter trailing behind. It is no wonder that leatherbacks are so seldom seen in the ocean. They are almost invisible to the serious as well as the casual observer.

The next day we went by Macklin's to thank George Mendonca for his help. We went into his office, and Chris pointed out a picture on the wall. It was the famous picture by Alfred Eisenstaedt from *Life* magazine of the sailor kissing the nurse in

Times Square on Victory in Japan Day in August 1945. I remember thinking, "It's George!" Even years later the resemblance was unmistakable. Of course, there was a big search to figure out who the sailor and nurse were, and many men have claimed to be that sailor. A Navy War College panel concluded, based on the evidence (scars, tattoos) that it was indeed George. In 2005 photo analysis expert and former Yale dean Richard M. Benson stated, "It is therefore my opinion, based upon a reasonable degree of certainty, that George Mendonsa [*sic*] is the sailor in Mr. Eisenstaedt's famous photograph." Looking at George and the photo, I simply had to agree. There was something unmistakable there.

George had a strip of pictures on the wall that had been taken on that day. Like any good photographer, Eisenstaedt had shot several pictures, and only one was in the magazine. The other pictures included one in which the right arm of the sailor was exposed, revealing his tattoo. Well, George showed us the same tattoo and even better, a picture of him taken before Victory in Japan Day, before the famous picture, showing the same tattoo. He was the guy alright. I asked him who was the nurse, and he said he had no idea. He just was so happy that the war was over, had had a few drinks, saw her, and gave her a big kiss. He never met her again.

The Dark Side of the Drift

Life is perilous for small sea turtles during their oceanic phase, and they face increasing odds against survival. The ocean is filled with the garbage of one of the turtles' biggest threats: man. The same currents and fronts that bring the little turtles together with logs, trees, seaweed, and other natural places to rest and hide also now bring all kinds of debris from human civilization, including the BP oil. You can walk along any beach in the world and find all kinds of interesting things. At Tortuguero a few years ago I made a collection of small items that washed up on the beach. They included a toy soldier, a dead balloon, plastic pellets about the size of a small pea, some larger pellets of Styrofoam, globs of oil, some rope, and a few other items.

An incredible amount of garbage floats around in exactly the places in the ocean where sea turtles live. This stuff includes plastic bottles, plastic bags, plastic sheets and tarps, beads, pellets, line, rope, strapping material, pieces of plastic chairs, monofilament fishing line, plastic six-pack yokes, old fishing nets, and every other kind of material that people throw away. If you throw trash out where water flows, it will eventually end up in the ocean, and it will eventually converge on the very place where the little loggerheads and other turtles spend their oceanic life. Since little sea turtles bite and swallow anything that they can get into their mouths, it is not surprising that they are often choked and poisoned by this garbage. The driftlines that used to give these turtles life are now driftlines of trash. The Pacific has some huge areas of garbage that stretch for many miles. How many little turtles are chowing down on that stuff? Too many.

A Plastic World

We all like plastic and have become dependent on it. It is everywhere in developed and undeveloped nations. I just took a look in the refrigerator. There was a plastic ketchup bottle, a plastic mustard bottle, the plastic yogurt container, a piece of meatloaf covered in clear plastic wrap, lettuce in a plastic bowl, carrots in a plastic box—the list goes on. Where does all that plastic go? Some we recycle, some we don't.

Some plastic gets buried in landfills or burned. Some ends up in rivers and makes its way to the sea. Rubber and plastic balloons filled with helium float away from birthday parties and graduation ceremonies at high schools and universities. Some are biodegradable, some are not. Some degrade quickly, some take too long. Many of them come down in the ocean. Drexel students who work in the Pine Barrens of New Jersey regularly come upon colorful dead balloons and collect them. Over a summer they can fill a garbage can with such waste from the sky. The rest of the balloons from places like Pittsburgh and Ohio float off into the Atlantic.

Underwater, a plastic bag looks like a jellyfish. It fooled me once and I know better. I thought I saw a jellyfish while snorkeling along the shore of St. Croix in the U.S. Virgin Islands. I swam up to take a picture, and as I got close I saw that it was a potato chip bag! Plastic and Styrofoam pellets look like little morsels to hatchlings, which seem to eat anything novel that they can get into their mouths. Therefore, many sea turtles eat plastic. The largest leatherback ever seen died in the sea off Wales in 1988 after being caught in a lobster line. Chris Luginbuhl sponsored a necropsy of the turtle at the Wales Museum. There is a great picture of the 3-foot-long plastic bag that they pulled out of the turtle's stomach. The lobster line killed the turtle (it got tangled and drowned), but the plastic can't have been good for it.

In 2000 on Playa Grande, Jenny Lux, a research assistant from New Zealand, and an Earthwatch volunteer were lying behind a leatherback, waiting for her to lay her eggs so that they could count them. The turtle was pushing and making all the right moves and actions, but nothing came out. That went on for a few minutes. Then Jenny saw that there was a little something sticking out of the cloaca, the common opening turtles have for both the reproductive and digestive tracts. It looked like it was clear. Jenny pulled on it and out came a 2-foot-long plastic bag covered in goop. Talk about gross. It really stank. Then the eggs came out, eighty-six in all. That turtle came back to nest seven times that year. Would the eggs have eventually come out? Who knows? But surely this is not the best we can do.

6

THE DEADLIEST CATCH: THE OTHER SIDE OF FISHING

As juvenile sea turtles approach the adult stage, their risk of death at the hands of humans increases because they tend to start inhabiting areas where fishermen work. Their dangerous encounters with us range from a loggerhead taking the bait of a recreational fisherman to leatherbacks drowning in commercial gill nets. These are but two of nearly a dozen deadly possibilities.

In 1990 a comprehensive review of sea turtle mortality in U.S. waters by the National Academy of Sciences concluded that the capture of sea turtles in U.S. shrimp trawls was the leading cause of death for these animals. The study reported that up to 50,000 loggerheads and 5,000 Kemp's ridleys were drowning each year. Those were incredible numbers, and the fishing industry disputed them, saying that they were overestimates. A political fight erupted that lasted many years and included dodging, deflecting, demurring, reflection, discussion, delay, cussing, and lying. Perhaps this is the way things always go (think of the legal debates over tobacco litigation), but for sea turtle conservationists valuable time was being lost. For shrimp fishermen, their "way" was threatened.

Shrimp Trawling: How Not to Solve a Problem

The U.S. National Marine Fisheries Service (NMFS) knew about the trawling problem for many years, since the 1970s at least. The federal agency began to develop a solution—modifications to shrimp trawls—in 1978. The goal was to find a device that would allow sea turtles to escape from a shrimp trawl through some sort of escape hatch while allowing shrimp fishermen to keep catching shrimp. The mythical win-win. In truth NMFS was required to do something, as the Endangered Species Act (passed by Congress and signed by President Nixon in 1973) mandated that the government agencies intervene in such cases. NMFS was not in a hurry, of course. Few federal agencies in the United States or elsewhere go looking for a fight with fishermen. Typically, they need to be prodded by conservation groups and the public.

By the early 1980s NMFS scientists had developed a Turtle Excluder Device (TED) that excluded 97 percent of sea turtles with negligible loss of shrimp. However, it was big and bulky and shrimp fishermen did not want to use it. In fact, most shrimpers did not want to use *any* TED and objected to the government telling them what to do. So NMFS delayed the requirement to use the TED and initially made its use voluntary. A few conscientious fishermen did employ them, but most did not. Meanwhile tens of thousands of turtles kept drowning.

During the 1980s, two people and one environmental organization entered the picture and changed the dynamics of the situation. First, a shrimp fisherman in Georgia, "Sinkey" Boone, brought to everyone's attention a grate-and-trapdoor device that he had been using for years to keep cannonball jellyfish out of his nets. It was lighter, smaller, and more effective than the device that NMFS had been pushing. It seemed just as effective with regard to saving sea turtles and keeping shrimp but was cheaper and easier to use.

Second, South Carolina state biologist Sally Murphy got into the act. Sally had graduated *cum laude* from Armstrong State College in Savannah, Georgia, with a bachelor of science degree in biology and a minor in chemistry, and later earned her master of science degree in biology from the University of South Carolina. She's down-to-earth, outgoing, and tough. She was hired as the sea turtle coordinator for the State of South Carolina and was also co-leader, along with Dr. Peter Pritchard, of the fledgling Marine Turtle Recovery Team. She pushed the use of TEDs throughout the southeast.

Third, an environmental organization called the Center for Marine Conservation (now the Ocean Conservancy) went after NMFS legally, pushing the service to require the devices in accordance with the Endangered Species Act.

Finally, in 1987 NMFS published regulations that required the use of Turtle Excluder Devices (usually called TEDs, though sometimes soft-pedaled as Trawler Efficiency Devices). The TEDs were to be attached to shrimp nets in the southeastern United States, however Sinkey Boone's device was not one of the approved models. Shrimp fishermen began screaming when they heard about the regulations. They pushed their state legislators, governors, congressmen, and senators. They demanded relief from the onerous regulations that they claimed would put them out of business. The states of North Carolina and Louisiana sued the federal government in order to stop the regulations. The U.S. Senate acted to delay implementation of the regulations. When the states lost their cases in federal court, Louisiana passed a state law stopping state officials from enforcing the federal law.

At this point Sally kicked into high gear. She thought that if Boone's device, called the Georgia Jumper, could be adopted as part of a regulation then shrimpers might go along with it, since it was designed by one of their own and not by the government. Shrimpers were not happy with Sally or Sinkey. He was ostracized by many, and she was condemned.

Congress continued to delay implementation of TEDs, and Louisiana sued again in 1989, further delaying action. By 1988 Sally had convinced South Carolina regulators to require TEDs, however, the state then passed a law whereby the South Carolina regulations could not be enforced unless the federal ones were in effect. More turtles kept drowning. Finally, in 1989 NMFS finalized regulations that required the use of TEDs only during May through August (the deadliest months) beginning in 1990.

Within days of the start of enforcement things fell apart. First the Coast Guard suspended the regulations to check to see if the TEDs got clogged with sea grass. Then shrimpers in Texas began a major protest. They blocked harbors, threatened people, and disrupted navigation. I remember watching the national news in disbelief as the Coast Guard stood by and let the shrimpers break the law in Galveston and other harbors. The Coast Guard apparently did not want a confrontation and neither, it seemed, did the NMFS. Then the unthinkable happened: the U.S. secretary of commerce, who oversees NMFS, suspended the regulations!

Environmental groups then sued in federal court. The judge ordered the secretary of commerce to issue TED regulations, effective immediately. However, the judge left a big loophole, one big enough to drag a shrimp net through. Shrimpers had to use TEDs unless they checked their nets every 105 minutes. This was supposed to let them retrieve any turtles from the nets and save them from drowning. It was known that sea turtles could hold their breath for about 30 minutes in warmer waters, so the loophole was very dangerous. In addition, law enforcement would have to time every net tow for every boat (quite impossible)—as opposed to pulling up to a boat and simply checking to make sure the gear was in place.

Environmental organizations went back to federal court, and strict regulations finally went into effect a short time later. The number of dead turtles washing up on shore dropped dramatically as soon as the regulations started being enforced. Also, many shrimpers in Georgia and the Carolinas reported they had less bycatch (unwanted fish, jellyfish, and trash that clogs nets). In the fall, when TEDs came off the nets, dead turtles started to wash up again.

Finally in 1994, after a 2-year delay, regulations required the use of TEDs year-round, from the Carolinas to Texas. Fishermen in Texas and Louisiana were still in a foul mood and protested but did not try a blockade. Within a few years everyone seemed to adapt and follow the law. Problem solved? Unfortunately, no.

There was no requirement specifying the size of the TED exit opening. When the devices were first approved in the late 1980s, the openings were large and any turtle could escape the net. Then shrimpers in many areas started to tie the openings smaller and smaller, arguing that they were losing shrimp. As a result, more and more big sea turtles were drowning and washing up on the beaches.

Sally Murphy led the charge of those complaining to NMFS that the openings were too small and sent data on the dead turtles that washed up on the beaches of South Carolina. Dave Harrington, a University of Georgia Marine Extension agent in Brunswick, Georgia, also complained. Sally and her volunteers measured 89 nesting loggerheads and found that most of those animals had body depths greater than 12 inches. Instead of adopting the obvious and simple solution of changing the size regulation, the NMFS set up a complicated system whereby if enough leatherbacks were present (10 in 50 miles of transect line

Hawksbill turtle (*Eretmochelys imbricata*) swimming over coral reef, Japan.
PHOTO COURTESY KATSUTOSHI ITO / NATURE PRODUCTION / MINDEN PICTURES.

Top, Boys on leatherback turtle in a 1910 postcard.

Bottom, Archie Carr with local fishermen on board the RV *Alpha Helix.* PHOTO COURTESY JAMES R. SPOTILA.

Loggerhead hatchlings on North Carolina beach.

PHOTO COURTESY BARBARA BERGWERF.

Fishermen loading green turtles into *cayuga* on Miskito Cays.
PHOTO COURTESY JAMES R. SPOTILA.

Top, East Pacific green turtle hooked on a longline in Costa Rica.

Bottom, East Pacific green turtle pulled from a longline in Costa Rica.

PHOTOS COURTESY SAMUEL FRIEDERICHS.

Top, Leatherback emerging at night on Playa Grande, Costa Rica.
PHOTO COURTESY JAMES R. SPOTILA.

Bottom, Lights from houses on Playa Grande light up the night.
PHOTO COURTESY JASON BRADLEY.

Researcher Gabi Blanco and the egg poacher Ricardo.

PHOTO COURTESY SAMUEL FRIEDERICHS.

Top, Leatherback with people on Playa Grande.

Left, Closeup of leatherback head shows cutting jaws and pink skin from blood flow.

PHOTOS COURTESY BARBARA BERGWERF.

observed from an airplane), then shrimping would be closed in the area out to 10 miles for 2 weeks except for boats that voluntarily used leatherback-sized TEDs.

I remember thinking that those regulations seemed awfully complicated and hard to implement and enforce. I wondered, "Why not just have all boats use a bigger TED?" I could not understand why the regulations were so convoluted, cumbersome, and ineffective. I remembered the mantra of many a good program—keep it simple. Other biologists could not understand the situation either, and no one could figure out who made that stupid rule and, more importantly, who could fix it. Sally kept complaining.

Finally, in 1999, Sheryan Epperly and Wendy Teas of the NMFS, two superb scientists, published a study of strandings data and concluded that the larger opening would save a lot of sea turtles and help to restore sea turtle populations. Officials within NMFS now had their own internal report, so everyone assumed the problem would soon be solved. Yet, it was not to be.

By this time Sally and her husband, Tom Murphy, were conducting flights to document leatherback abundance. When they found their ten turtles, the NMFS implemented its complicated closure system. If the airplane broke down, leatherbacks washed up on the beaches. One year newspapers reported close to a hundred dead leatherbacks washed up on the beaches of Georgia. In another case, many leatherbacks washed up in Florida. The plan just wasn't working. South Carolina finally took action, passing a law requiring large-sized TEDs on all shrimp nets in state waters. With that political cover, NMFS followed with a similar federal requirement from North Carolina to northern Florida for a period of 30 days. Then in August 2003 the NMFS finally required the larger opening for all shrimp nets all along the East and Gulf coasts of the United States. It took almost 30 years from the time the problem was first identified until a

proper solution was implemented. It might have been 40, or never, if not for the efforts of Sally, Sinkey, and others who kept their eye on the prize.

International Shrimping

The success story in the United States does not mean that turtles are safe from fisheries. Far from it. Most countries do not force fishermen to use TEDs, from small-scale fishing by local people in Africa to large industrial trawlers from China, Taiwan, Spain, and other countries in the European Union. There are dramatic impacts from such trawling. In the 2008–2009 leatherback and green turtle nesting season, there was a dramatic drop in the number of nesting females coming ashore on Equatorial Guinea's Bioko Island and in Gabon and other local beaches. Rumor had it that a Chinese trawling fleet had worked the Gulf of Gabon a few months before. Local fishermen stopped catching any fish, and biologists were left to wonder if the fishing fleet had vacuumed up their turtles along with the fish.

In West Africa there is no enforcement of laws related to trawling, so industrial fleets often fish in nearshore areas that are supposed to be reserved for local fishermen. They also drag their nets through so-called marine reserves. To make matters worse, countries often negotiate away environmental enforcement. Such is the case in Central America.

In Costa Rica, Taiwan "donated" funds and built a bridge over the Tempisque River connecting Puntarenas Province with the Nicoya Peninsula. The Friendship Bridge made the trip to the nesting beaches of Guanacaste from the capital of San José in the Central Valley much more convenient. However, it came with a price. Shortly thereafter the Taiwanese fishing fleet began to unload its catch at the docks of Puntarenas (in fact their catch was shark fins, which involves cutting off the fins and releasing the doomed shark—an illegal practice in Costa Rica). Conservationists, led by Randall Arauz and his PRETOMA organization, conducted numerous protests and filed lawsuits in the Costa Rican Constitutional Court.

Eventually Costa Rica became the first nation in Central America to establish diplomatic relations with China, and Taiwan moved its fleet away to land its catch at ports in El Salvador. Taiwan was not very happy with the disloyalty shown to it by the government of Costa Rica's president, Óscar Arias. China thanked Costa Rica with a large economic development grant of $500 million, and the Chinese fishing fleet is expected to arrive in Puntarenas soon.

In India, tens of thousands of olive ridley turtles have been killed each year since the mid-1980s by the local and industrial shrimp trawling industries in the vicinity of the great olive ridley *arribada* beaches at Gahirmatha, Rushikulya, and the Devi River mouth along the Orissa coast. India has resisted international pressure to use TEDs, and the nation's fishing effort has greatly increased during that time. Thailand also catches many sea turtles in its shrimp nets and refuses to use TEDs. Likewise, Malaysia and Indonesia employ many shrimp trawls, do not use TEDs, and drown uncounted numbers of sea turtles, including green turtles, leatherbacks, and olive ridleys. The bottom line here is simple: if you eat shrimp, buy American, because those shrimp are caught by boats using TEDs. If you don't know the origin of the shrimp, have pasta instead. Otherwise, to twist an old quote, it's not shrimp you're buying, it's turtle lives.

Longlines of Death

In the old days, defined as a time before we were born—say in the 1930s or 1940s—fishermen caught swordfish the old-fashioned way. They went out in a boat and harpooned big swordfish as they were swimming near the surface, usually on a calm day so the fishermen could see the big dorsal fin breaking the surface of the ocean. One man stood in a tall crow's nest above the cockpit of the boat and spotted the fish. He directed the captain to motor over and put the bow of the boat just behind the swordfish. A really skilled fisherman, the striker, stood in a bowsprit 15 to 20 feet out in front of the bow of the boat with a harpoon.

As the bowsprit came over the fish, the striker threw the harpoon into the back just near the dorsal fin. If the hit was clean it drove the harpoon tip deep into the back of the fish. The point detached from the wooden shaft and as the swordfish dove to depth, as if to reach safety, it pulled out several hundred feet of thick line attached to a buoy from a big bucket or bin on deck. The boat would follow the buoy. After some hours the fish would tire and the fishermen would motor over and reel it in. Once they hoisted the fish on deck they would dispatch it with a quick slice of a big knife behind the gills.

In the old days, people caught tuna by going out in a boat and "fishing the birds." They looked for birds diving into the water after small fish chased to the surface by tuna. Then they got over the tuna and used big wooden poles to drop lines with large hooks over the side. The tuna attacked anything shiny and grabbed onto the hooks. The men then lifted the poles and the tuna came with them, over the side, and fell into a big box, or the hold, containing ice. To improve that method the fishermen would surround a school of tuna with a net and corral them until they could catch all of them with their hooks.

I remember sitting in my living room and watching a fishing program on TV one Saturday afternoon when I was eleven or twelve. The program showed a boat with twenty big, strong men, each with a pole. Each would dip his pole into the seething mass of tuna and lift it up over his head, launching a 3-foot-long tuna into the air and into the fish box. It was amazing: the fish came off the hook and was airborne, so the hook must not have had a barb. No sooner was the fish gone overhead than the guy was dipping the hook back into the water. For some really big fish, probably bluefin tuna, two or even three men would hold poles connected to one hook. Then a huge 6- to 8-foot fish would come up over the side and fly into the hold. Now that was the way to fish! I decided that when I grew up I would become a fisherman. Times changed.

Fishermen got even more efficient and used airplanes to spot schools of tuna or locations where big swordfish dwelled. They understood the currents, fronts, and temperature shifts and got better and better at finding fish. One invention that drastically improved efficiency was monofilament fishing line. Someone thought of tying hooks to lengths of monofilament line with floats and putting them onto even longer leaders that could be set out from boats. The longline was born.

Gone was the boat full of strong men. It was now manned by a crew of four to six who could catch an incredible number of fish by laying longlines in a location where the currents were right. Soon the longlines were joined by another method, giant nets called purse seines. Tuna, swordfish, marlin, sharks, mahi-mahi—these were all getting caught in greater and greater numbers.

Today there are at least 1,400,000,000 longline hooks put into the world's oceans each year. That is a lot of hooks! They are spread out on lines that are each 100 km (60 miles) long, with thousands of hooks baited with squid or fish. Every so often lightsticks are added to attract fish. Lines are set at the surface for swordfish and deeper for tuna.

I woke up early one morning at Playa Grande, Costa Rica. There had been a storm the night before. Gazing down the shore it looked to me like there was a big boat beached about halfway down. Several of us walked over, and there it was, a Costa Rican longline boat about 60 feet long, high and dry on the beach at low tide. The story was that the boat was steaming back from Cocos Island and the engine broke down. It was only 20 km (12 miles) from Playa del Coco, its home port, but washed up on Playa Grande in the storm. The crew were all glad to be alive but unhappy to be beached in a national park.

Rotney Piedra, the park director, and one of his rangers boarded the boat to see what it was carrying. It was full of fish. That morning, trucks full of ice arrived behind the beach and the fishermen began to unload their catch. It would have spoiled if they did not get it to market quickly. One big, green fish after another was hoisted off the boat. Some fish were dragging on the ground. I remember thinking, "These longlines catch a lot of fish." Later that day the tide came in and another boat showed up to pull the fishing boat off the beach. Reading reports of how many fish longlines catch is helpful, but seeing one being offloaded leaves a deeper impression.

Unfortunately, longlines do not just catch fish. They also catch birds, like albatrosses, and dolphins and seals. They also catch sea turtles. Loggerheads, olive and Kemp's ridleys, and greens go for the bait and get hooked in the mouth. Even if they are freed from the hook, they may die from the wound, especially if the hook has impaled the eye or gone into the gut. Leatherbacks appear to be attracted by the lights, and perhaps when they try to eat jellyfish that get stuck on the lines. The lines get tangled around or hooked in their flippers. Occasionally leatherbacks are hooked in the mouth, even though they are not supposed to eat the fish and squid that are put on as bait. Andres Domingo and colleagues in Uruguay put observers on longline boats in the South Atlantic. They found that, of 3,149,638 longline hooks set, the number of turtles caught were as follows: 1,128 loggerheads, 259 leatherbacks, 29 olive ridleys,

28 green turtles, and 42 turtles that were not identified. That is a lot of turtles!

Even if the hook is not too deep, a turtle caught on longlines has to be caught in a way that it can swim to the surface to breathe. Then a fisherman has to remove the hook properly before letting the turtle go. Sometimes they do. If the hook is buried in the flipper, some fishermen simply cut off the flipper to get the hook back. Hooks are expensive and turtles are worthless in many a fisherman's eyes.

Soon after obtaining her PhD, Rebecca Lewison was hired for a time by Duke University, where she worked with Larry Crowder on an interesting project in which they tried to determine how many turtles got caught on longlines each year. They combed the scientific literature for articles, talked to fisheries biologists, and requested data from national and international regulatory agencies and associations. Many of these sources said that they did not have such data. Others said that they would cooperate but never sent information. After a great effort and a lot of analysis, Rebecca's team came up with some excellent information.

Based on data from the Atlantic, Pacific, and Indian oceans, they estimated that one turtle was caught per every 5,600 hooks. That translates to catching 200,000 loggerheads and 50,000 leatherbacks in a single year. In the Pacific they estimated that 30,000 loggerheads and 20,000 leatherbacks were caught in the year 2000. If correct, that means that 45 percent of the loggerheads and 63 percent of the leatherbacks in the Pacific Ocean were caught on longlines that year. Of those turtles, as many as 6,000 loggerheads and 3,200 leatherbacks were killed. Those numbers are pretty striking when we realize that the mortality rate for these turtles from longlines is about 10 percent a year. That is clearly unsustainable. Those fishermen catch lots of sea turtles. Something needs to be done fast.

So what has been the response of fishermen? They have disputed the facts, disparaged the study, and delayed any change in the way that they fish. They claimed the science was sloppy, that Lewison did not use the best data (the data that the fisheries people wouldn't give her), and that the estimates of turtles caught were too high. However, the Lewison study and the Domingo study in the South Atlantic both point to about the same impact.

What has been the response of governments? In most countries there has been no response. In the United States NMFS has dragged its feet. Lawsuits filed by environmental and conservation organizations led the federal courts to close the Hawaiian longline fishery over the objections of NMFS, which has since been looking for ways to reopen the fishery. The best idea they have come up with is circle hooks.

Normal hooks are in the shape of a J and are very effective in hooking a fish—or a sea turtle—in the mouth, gut, or elsewhere. The problem is that they are hard to remove and cause a lot of harm. Circle hooks are just that: they are in the shape of a circle with a small lip at the end that turns in. The hook tends to get caught in the mouth and not the inside of the fish or turtle.

Early tests in the Atlantic with circle hooks have given some positive signs. Fewer turtles were caught, and fewer of those badly harmed. Then tests conducted in Costa Rica indicated that the catch rate was just as bad as with J hooks and that there were a lot of injuries to turtles. So the jury is out at this time. Meanwhile, instead of making the protection of sea turtles its priority, NMFS continues to make catching fish the priority. It is a *fisheries* service after all. Sea turtles, those animals that the Endangered Species Act states the NMFS must protect, still come second.

There is some hope in this story, but not much. The World Wildlife Fund has been carrying out tests with circle hooks, working with local fishermen in the eastern Pacific in cooperation with the Inter-American Tropical Tuna Commission (IATTC) and other partners such as the Ocean Conservancy and NMFS, which provided free hooks, some funding, and scientific advice. These tests in Ecuador, Panama, Costa Rica, Colombia, Peru, Nicaragua, El Salvador, Guatemala, and Mexico indicate that circle hooks catch fewer turtles and more or similar numbers of tuna and swordfish than J hooks. However, J hooks catch more mahi-mahi than circle hooks. Sea turtles are still being caught, but they are being released more often and in better condition. Of course, changing a hook does not change the problem of getting tangled in the line.

Nevertheless, even this program has limitations. Two of the guiding principles are that fishermen continue to fish and that the program is voluntary. Voluntary programs did not work with TEDs in the United States, and they won't work with fishermen in other places, either. Voluntary regulation of fishing works about as well as voluntary regulation of speed limits for drivers. The real issue is that with sea turtle mortality rates still way beyond sustainable levels in leatherbacks and loggerheads in the Pacific Ocean, a program that takes decades to implement may well result in the extinction of leatherbacks and loggerheads in the Pacific.

There are enough scientific data to know where the turtles go in the ocean. Marine reserves—areas set up as no-fishing zones—can protect sea turtles. A program that combines establishment of marine reserves, closures to fishing of select nonreserve areas during sea turtle migration, and the mandated use of circle hooks could prevent the extinction of loggerhead and leatherback turtles in the Pacific Ocean.

Longlines and shrimp trawls are two of about a dozen problems sea turtles face in the oceans. Throw in our earlier discussion of plastics, and we've covered the first three. The next chapter will delve into a few more.

7

OUT OF THE FIRE: THE GAUNTLET CONTINUES

There is a story sometimes told by sea turtle biologists of a researcher sitting in the kitchen of an old-time fisherman. The fisherman was very curious about the life cycle of sea turtles, having caught them by accident his whole life using a variety of gear. As the researcher explained all the travels a sea turtle undertakes and perils it faces in a lifetime, the fisherman famously shouted, "They gots to run a gauntlet of fire!" Indeed they do.

At Playa Grande in Costa Rica our team calculated that the mortality rate for adult female leatherbacks was about 20 percent per year. Pelagic longlines are responsible for half of that mortality. What causes the other half? The answer is almost certainly gill nets. A gill net is a mesh of thin, but strong, monofilament line. It is very hard to see underwater, even from an inch away. A fish swims along and sticks its head into the net. It moves forward until its head gets stuck. If it has moved past its operculum (gill cover) it cannot move backward, as the line slips under the operculum. It is caught. Once a sea turtle gets its head or flipper tangled, the clock starts. The turtle has to get free before it needs its next breath. It has 30–45 minutes to either get free or hope the fisherman comes. Many fishermen leave their nets for many hours, so the turtle better get free or it will die.

Things are actually better than they were in the 1980s. Back then industrial fishermen in large vessels set out oceanic drift nets that were 200 meters deep and 100 miles long. Those walls of death killed everything that came their way. Eventually, through the efforts of the United Nations (yes, it does some worthy things) those nets were banned. But better is not well.

There are still some big drift gill nets out in the oceans. They are called pirate nets because no one claims them as their own. Who sets them? What countries are involved? How many of them are there? Those questions go unanswered. Even without those nets, however, the problem of gill nets is huge. The smaller ones kill many turtles in the Mediterranean Sea. In Trinidad, gill nets routinely catch nesting leatherbacks. Joseph Knorr, a biologist from Canada, calculated that coastal gill nets along the Pacific areas of Central and South America kill at least 10 percent of leatherbacks each year.

Many of the problems with gill nets are attributed to the small, unregulated local fishermen in small boats that range in size from small canoe-like *cayugas* or *pangas* to fishing boats that are 50 feet or so in length. Fishing in this way is common in West Africa, Bangladesh, India, Sri Lanka, all over Indonesia and the Philippines, in Tanzania, Eritrea, and just about every other country that has a coastline, including the United States. Fishermen lay out the nets across the ocean floor and along the surface for miles and miles and then come back the next day to check them and collect their fish.

Can you imagine what it must be like to be a sea turtle that swims into a gill net? Here you are, swimming along slowly with broad sweeps of your flippers, your head down looking for crabs and clams and such along the bottom. All of a sudden you bump into something in front of you. You can't quite see what it is. It is not hard, so it is not a rock or a piece of wood. What is it? You take a sweep with your front flipper to move it out of the way and find that your claw halfway down your front flipper is tangled. Algae? It will soon break free. You push ahead trying to get through it and get your head stuck in between the netting. A few more swipes and both flippers are entangled. So you do what a turtle does. You keep swimming forward and drag the net along.

Soon you are being held back and you are running out of oxygen. You need to breathe and so you start upward. But you can't quite reach the surface. The net is holding fast. Desperation sets in. Your turtle mind is getting a little foggy. You push with all your strength for the surface. Your muscles begin to fatigue as the lack of oxygen is reaching a critical level. Lactic acid builds up and you need to rest. After a moment you try again. You are bobbing up and down, perhaps a meter or two from the air.

Later, the fishermen come by to check the net. Damn, another turtle tore up their net! They cut you loose to fall to the bottom or bring you back to sell in the local market, all the while complaining that they will have to mend the net. Either way there is one less turtle in the ocean. If there is a scientist around doing a survey of incidental catch in the local fishery, he will try to determine if you drowned or were simply asphyxiated. As if it makes a difference.

It is interesting that in the United States, which prides itself as being the most advanced nation in the world, we still allow coastal gill-netting on both the East and West coasts. The California coastal gill-net swordfish fishery was closed because of the leatherback turtles in 2007. In 2009 the NMFS proposed to open it again. So it goes; NMFS, an agency that is supposed to protect sea turtles is often called the fox agency—the fox guarding the hen house agency. Their mission: protect endangered species while promoting fishing. It is pretty hard to serve two masters.

Other Ways to Kill a Turtle

How many other ways can we find to kill off sea turtles? One is scallop dredging. In that fishery, boats drag a big, heavy metal cage that scrapes along the bottom and dredges up all of the shellfish in the sand or mud, along with literally tons of skates, fish, sharks, crabs, lobsters, and once in a while a sea turtle. There is a lot of scallop dredging along the East Coast of the United States, often in the northern states such as New York. Scallop dredges catch and crush sea turtles, especially loggerheads that are cruising along the bottom and munching on those same scallops.

It probably won't surprise you when I reveal that NMFS permits that industry to kill a certain number of sea turtles each year. It has known about the "loggerhead problem" for many years but continues to publish opinions supporting a continuation of the practice. The fishermen proposed to put chains on the front of the dredges to divert the turtles, sort of a turtle deflector device, or TDD. No one knows if this idea would work. The debate continues.

About the only fishing method that does not routinely kill sea turtles is the pound net, or trap-net fishery, at least the way it is practiced in some areas. In a pound or trap net a long net leader runs from shore out to a funnel or set of boxes that eventually lead the fish into a big net with a small opening. The big net is a five-sided box, if you will. There is net on four sides and the bottom. The fish is funneled into the box through a small opening. It could swim out, but it would have to find its way through the maze. The top is usually open air (some have a net over the top, above the water, to keep birds out). When the fishermen check that net, they pull up in a small boat and hand pull the net up off the bottom. The fish are contained in the box and the fishermen can reach in with big dip nets and pick out the fish that are economically valuable. The other fish can be let go. The same is true for sea turtles, which are generally swimming around with the fish and can be picked up and released. This type of fishing is much less harmful to sea turtles than any other. There have been claims that some pound nets use wider mesh and that turtles have gotten stuck in these nets. If true, smaller mesh size should be mandated.

Dredging of harbors and ship channels also kills sea turtles, as they are vacuumed up by suction dredges. The agencies that monitor that activity actually put observers on board dredges to figure out how many dead turtles are represented by the bits of shell and flipper they can see on the surface of the dredge material. If they add up to too many turtles the dredging has to stop.

Boats slice turtles with their propellers and crush them when they hit them with their hulls. The only way to solve this problem is to keep boats away from turtles by restricting their use and speed. In many areas there are simply more boats than should be allowed. Florida and the waters around New York, for example, would look like a sea of buzz saws to a turtle.

Finally, some power plants eat turtles. Turtles can be sucked up by intake systems as huge amounts of water are needed for cooling. In 2006 one nuclear power plant, Oyster Creek, reported that it had taken in eight Kemp's ridleys, five alive and three dead. The NMFS decided that it could not determine if those dead turtles died in the intake or were dead before they were sucked up! And there were not even fishermen to protect.

Conflict of Interest

The conflict of interest that is apparent in the United States with regard to NMFS is not unique. In Costa Rica, INCOPESCA, the fisheries agency, also drags its feet in enforcing the TED regulations that are required by law. INCOPESCA has even warned shrimp fishermen before inspections so that they could take out the TEDs from storage and put them on their nets. That is, until a biologist in PRETOMA took pictures and filed a report with the U.S. Embassy, which required the United States to decertify Costa Rica as having TED-safe shrimp. Costa Rica was then prevented from exporting shrimp to the United States.

In the United States, the NMFS has allowed sea turtle protection to take a backseat to fisheries promotion. That was true with the development and implementation of TEDs, longline fishing, coastal gill-net fishing, scallop dredging, flounder trawling, and the operation of power plants. The economic activity has to go on, and sea turtle protection has to be done only as long as it is not inconvenient. It is virtually impossible for an agency that has as its primary purpose the promotion of fisheries to also regulate those fisheries and protect endangered species. That was proved long ago with nuclear power.

In the 1950s the Atomic Energy Commission (AEC) had the task of promoting the peaceful use of nuclear power and of regulating the nuclear power industry. It soon became apparent that it could not do both. Congress formed the Nuclear Regulatory Commission (NRC) and gave it the responsibility of regulating nuclear power. It still does so to this day. It is not perfect, but it does a much better job than the AEC did in days gone by.

Is it necessary to remove the regulatory function for endangered species from the NMFS, which is housed within the Department of Commerce (yes, the Department of Commerce!)

and give that function to a federal agency that has a better record of protecting species, such as the U.S. Fish and Wildlife Service? Do we need a new marine protection agency? Can the good people in NMFS (there are some great conservationists there) be effective when they have to report to managers who see their job as promoting fishing and who themselves report to the secretary of the Department of Commerce? I'm dubious.

8

RETURN TO THE BEACH: YOU CAN'T GO HOME ANYMORE

Sea turtles grow up in the ocean. When they mature, they migrate hundreds to thousands of miles from their feeding grounds to their mating grounds and nesting beaches. They face many dangers from fishermen and garbage in the oceans and then return to a nesting beach that usually is the one on which they originally hatched.

Depending on the species, between 11 and 30 or more years might have passed since the turtle hatched from an egg on that beach. For females returning to nest, a lot may have changed. When she left as a hatchling, the beach may have been dark. Now, she may find a new hotel, or houses, or condos. Electricity lights up the night. Noise might blare from bars built directly on the beach. Will she recognize her natal home? Will the water smell the same as when she left? Will she be disoriented by lights? Will she come ashore to nest? Or will she, as sometimes happens, discharge her eggs at sea?

These are the issues and questions that face us when we consider the situation on nesting beaches today. Try going back to *your* old neighborhood. Maybe you wandered in the local woods when you were a child or fished in a stream or pond when you were a teenager. Now the woods may be a housing development or a series of huge McMansions scattered on big lots. The stream may be brown. The trees on your Maple Street may be gone. You can't go home, and, so it seems, neither can many sea turtles.

The only places in the United States that haven't changed much are national parks, state parks, national wildlife refuges, and military bases. Surprised by the last one? You will find more wildlife and endangered species on army and other military installations than almost anywhere else in the United States. I learned that when I filled a post in the Clinton Administration—as chief environmental scientist for the Army. I even had an office in the Pentagon. I learned a lot from that experience. The Army has preserved 20 million acres of land that are home to 199 protected species. The other services, although they have less land, have done likewise. Part of what helps the species may be the guns. People tend to stay away, and the animals and plants thrive on the open land that serves as a buffer for training areas. I have often wished that there were more military bases near the coast!

Beaches Long Ago

In the past, say 50,000 years ago, there were plenty of nesting beaches for sea turtles. Humans had not yet reached North America, and elsewhere they had had little impact on sea turtles. As ocean levels rose and fell and storms washed away or built up beaches along the coast, sea turtles had plenty of options for nesting sites. Along what is now the U.S. coast, individual loggerheads undoubtedly returned to the same beach from which they hatched, or to one close by. If their natal beach was washed away or had changed in the 25 years since they left it as a hatchling, they simply moved to another nearby beach. So sea turtle populations were controlled by their reproductive rate, the productivity of the oceans in which they swam, and predation on land and in the water.

Probably the biggest controlling factor was predation on eggs and hatchlings by birds and mammals on the beach. In many years, very few hatchlings may have successfully made it to the water to swim out to sea. But that did not matter, because the nesting females, the mother turtles, lived long and fruitful lives, returning to lay eggs multiple times. All they needed was one good year without predation or with greatly reduced predation and they could flood the ocean with their offspring.

Sea turtles, like freshwater turtles, appear to have evolved in the face of serious predation. As long as beaches could adjust to changes in the ocean, sea turtles had enough behavioral flexibility to adjust where they laid their eggs. As long as natural populations of predators waxed and waned on the beach, the production of hatchlings also waxed and waned. Sea turtle populations were in balance with their natural and biotic environment.

Reflections of Yesterday

We get hints of the past on some sea turtle nesting beaches today. When we first began studying leatherback turtles on Playa Grande in Costa Rica in 1990 there were no coatis, raccoons, or bird predators on the beach at night. The many years of human predation of almost all of the eggs probably discouraged those animals from coming onto the beach (both because of the presence of humans and the lack of eggs). Then by the mid-1990s, egg poaching had stopped and we began to see hatchlings.

Shortly thereafter, a raccoon started to come out to the beach at night near where a housing development had started. She appeared to live near the houses where food was available in the trash. She dug up olive ridley nests and hunted for leatherback hatchlings. She did very well indeed. We tried to chase her off the beach, but she seemed to just laugh at us as she scampered back to the forest only to return as soon as we moved on down the shore on our patrol. After a year or two she began to bring her offspring to the beach and the numbers of raccoons increased. They started to come to our hatchery and to climb over the fence to get at the hatchlings inside their little cages above the artificial nests. Volunteers guarding the hatchery would just scare away the raccoons with a flashlight. But the little animals did get to be a nuisance, and we could not leave the hatchery unguarded day or night.

Then one year the raccoons were gone. Maybe some of the construction workers ate them, or maybe they got mange. In any case we haven't had raccoons on the beach for a few years now. There are coatis on the roof of our biological station (they eat the papayas on our tree just before we pick them) but none on the beach. I suppose the new residents have not learned to eat eggs or hatchlings yet. Predation on a sea turtle beach follows the same wax-and-wane cycle as does the freshwater turtle nesting on the University of Michigan's E. S. George Reserve. Some years are good and some are not. A good year for turtles is a bad year for raccoons, and vice versa.

Olive ridleys nest in huge *arribadas* today. Kemp's ridleys nest in smaller *arribadas* because there are only a few thousand of them left in the Gulf of Mexico. Both species have evolved a nesting behavior that overwhelms the actions of predators of their eggs and hatchlings. At Playa Nancite in Santa Rosa Park, Costa Rica, you can see coatis digging up olive ridley nests, vultures sitting nearby waiting to get some eggs, coyotes coming down at night to eat both adult ridleys and hatchlings, and the resident crocodile and her offspring (see chapter 2) walking the beach and waiting in the ocean to catch adult ridleys and hatchlings. Frigate birds congregate to swoop up hatchlings during the emergence time, as do crested caracaras, yellow-crowned night herons, and other birds. It is a big all-you-can-eat buffet when the hatchlings emerge.

The cooperative arrangement between coatis and vultures is quite amusing. Symbiosis is the general name for a situation in which two different species live together in a close relationship. In this case, that relationship is beneficial to both species and is called mutualism. Coatis can smell out the nests and dig them up very well. But they are nearsighted and cannot see a predator creeping up on them when they have their head buried in the sand. The vultures can see very well but cannot dig. This mutualistic relationship works because the vultures essentially stand guard over the coatis as they dig, and squawk and fly away when a person or other large animal comes by. Then the coati picks up his head looks around and takes off running to the nearest tree. The vulture gets to eat the eggs that the coati leaves behind.

At Tortuguero long ago, thousands of green turtles used to come ashore to nest. But once people discovered the 20 miles of turtle heaven, they began to hunt turtles and their eggs, and greatly reduced their numbers. When Archie Carr first went there in the 1950s, he recounted the presence of jaguars on the beach hunting green turtles. Ocelots and other animals hunted there, too, along with people. Archie was so concerned that the green turtle population would disappear that he began Operation Green Turtle with the U.S. Navy, spreading hatchlings around the Caribbean to forestall the end of the green turtle in that part of the world.

Over the next few years the numbers of nesting turtles declined despite the presence of scientists on the beach. However, by 1978, when I first came to the beach, numbers of green turtles were reviving, owing to almost 30 years of study and protection.

That was a pretty good accomplishment, considering that it takes a good 25 years or more for a green turtle to grow from hatchling to adult and that there was still a legal hunt for green turtles out of Limón, the port city down the coast.

Today the protection and study have really paid off. There are more than 20,000 green turtles nesting on the black sands of Tortuguero each season, and the jaguars are back. Several jaguars now visit the beach and capture adult green turtles. You do have to be a little careful, however, as a jaguar might actually eat you.

One jaguar visits the beach at Playa Grande, and my student David Reynolds met it one morning in 1997. Dave was studying the metabolic rates of leatherback nests in the hatchery, but like all of the research assistants he took his turn doing the morning walk. You get up at the first sign of light, about 5 a.m., and walk the entire length of Playa Grande and Playa Ventanas, counting the number of nests laid the night before. We wanted to be sure we did not miss them on night patrol and to see if any of the earlier clutches hatched that night.

When Dave got down to the end of Playa Ventanas, he had the feeling that someone was on the beach behind him, watching him. He turned around and there was a jaguar staring out from the edge of the beach in the tall grass and bushes. Dave froze, as did the jaguar. They were both terrified, Dave for obvious reasons and the jaguar because it survives only by avoiding people. It probably lived on Cerro El Moro, the forest-covered hill at the north end of Las Baulas Park, and made its living hunting the overgrown cattle pastures and small wood lots scattered behind the beaches.

The next moment Dave was running back to the lab in one direction and the jaguar was running into the woods in the other. Now we do the morning walk in pairs. We have seen the jaguar again, once on the beach and once as it crossed the road between our lab and Enrique Chacon's restaurant up the hill. We try not to walk alone at night. Maybe, someone suggested, the jaguar ate the raccoons.

Imagine how it must have been, say on a beach in Florida. Jupiter Island will do. Thousands of loggerhead nests, raccoons, mountain lions, opossums, skunks, vultures, hawks, eagles, and other wildlife milling about waiting their turn to get into the bounty of life brought from the sea up onto the land by those nesting sea turtles. But there are so many turtles that the predators can't possibly eat them all. There was a whole food web based on sea turtles nesting along the coast of the eastern United States, all along the Caribbean Sea, and along many other shores around the world. That flush of energy reserves probably was just the thing needed to sustain a vibrant coastal ecosystem. Since adult sea turtles had few natural predators (big tiger sharks will eat them perhaps), they lived to a ripe old age, laying thousands and thousands of eggs in a lifetime. It was paradise.

Beaches Today

Today in many areas of Florida the turtles' beaches have been taken over by development. Houses, hotels, and high-rise apartments and condominiums dominate the shoreline. Sea walls, rock revetments, and sand-tube structures (rubber or plastic tubes 6 to 12 feet high and filled with sand) obstruct the beaches. Sea wall and other types of beach armoring prevent sea turtles from reaching the upper part of the beach, where they prefer to nest. In some cases sea turtles cannot get up out of the tidal zone to find a place to lay their eggs, and on urbanized beaches the lights disorient adults, and especially hatchlings, at night.

After a series of hurricanes in 2004, Florida's then-governor Jeb Bush waived environmental requirements and allowed more sand-tubes to be placed on beaches to "prevent erosion and stabilize the beaches." That took even more sea turtle nesting habitat away. For example, Jupiter Island is partially developed and losing sand from both the developed and undeveloped areas, owing to the presence of an inlet jetty that robs sand that normally would be deposited on the beaches from the long shore current. The beaches in the undeveloped areas gain sand from the dunes and still provide nesting sites. However, the developed stretches are armored; waves wash up to the walls and revetments, leaving no place for the turtles to nest.

It must be quite a shock for a loggerhead turtle to run right up against a sea wall when she tries to nest. Luckily, the Archie Carr National Wildlife Refuge on the Atlantic coast protects some of the best loggerhead and green turtle nesting in Florida, as does Cape Canaveral, the site of rocket launches. There are other places, each critical, all rare. Whatever is left out there is in urgent need of protection. Not just in the United States but around the world. The clock is ticking.

The number of loggerhead nests in Florida has been declining since 1998, when there were about 60,000 nests on the main beaches that are used to monitor the population. In 2009 there were 32,000, a 49 percent decline. Of course this is an echo, reflecting things that happened perhaps 20–25 years ago. What? Probably longlines, drowning in shrimp nets (before TEDs with proper-sized openings were fully employed in 2003), and drowning in other fisheries (e.g., in gill nets). Entanglement in marine debris, including lost or ghost nets, lobster pot lines, and plastic materials as well as ingestion of plastics and other nondegradable materials may also affect the turtles. Because loggerheads eat high on the food chain and toxins build up at each step in that chain, poisonous contaminants such as PCBs, mercury, and pesticides may have reduced their numbers. It is a bit puzzling that both green and leatherback nests are increasing in numbers on Florida beaches even as loggerhead nests decline. It may be that fisheries have a differential effect on loggerheads (perhaps shrimp trawling was particularly hard on them).

Traveling north from Florida you will find many fine sea turtle nesting beaches in Georgia and the Carolinas. Some beaches on the outer banks of the Carolinas and Georgia are protected by parks, state and federal wildlife refuges, or private owners. Those island beaches still have sea grasses, live oak forests, red cedar trees, and a natural vegetative understory (the layer between the canopy and the forest floor). The dunes are stable, and sea turtles going there see pretty much the same beach as they did when they left as hatchlings. Pinckney and Tybee islands in South Carolina and Wassaw, Blackbeard, and Wolf islands in Georgia are federal wildlife reserves. Georgia's Cumberland and Little Cumberland islands comprise the Cumberland Island National Seashore.

Other islands such as Kiawah Island in South Carolina continue to protect their beaches. Kiawah's township form of government maintains its natural dune system behind the open, unfettered beach and supports a modest but important 150 nests a year. In other areas, houses and hotels cover the beaches. For example, Jekyll Island, Georgia, was developed many years ago, and much of its beachfront is armored with rock breakwalls. Over the last 20 years it has had less than half the number of loggerhead nests as on protected islands such as Cumberland and Blackbeard islands.

Fortunately, the folks on Jekyll have seen the light. In their new redevelopment plans they seek to become a "model conservation community." They are keeping the open areas of the beach open and have passed an ordinance that prohibits all lights from shining out onto the nesting and non-nesting beaches from May 1 to October 31. That is a great improvement and good news for the turtles. In addition, Jekyll Island now hosts the Georgia Sea Turtle Center, which includes a rehabilitation center and veterinary clinic. Loggerheads coming back to nest there will find a more turtle-friendly environment, and we should expect more nests in the future. On your morning beach walk you can watch the center's turtle team excavate loggerhead nests to discover how many eggs hatched and how many hatchlings emerged from the nest. That is a very positive educational effort. If things continue in the same way Jekyll Island will serve as a model for other communities that want to help their sea turtles.

A Sea Turtle Hero

One of the developed yet still beautiful islands along the South Carolina coast is Hilton Head Island. It has 250 restaurants, 24 golf courses, 200 stores, 21 hotels, and many condominiums. However, it also has 50 miles of nature trails and bike paths and a new lighting ordinance that prohibits bright streetlights and prevents lights from shining on the beach. The Coastal Discovery Museum has a Sea Turtle Protection Project that patrols the beaches from May until October. The turtle team there relocates nests that are in danger of being washed away by the tides and protects the nests during the incubation period. This is another great improvement and good news for the turtles. Of course it helps to have a sea turtle person involved in the development of such measures and in the planning of such communities. Hilton Head was fortunate to have a real sea turtle hero in the middle of its turtle-friendly revival: Ed Drane, its architect and the island's urban planner.

Ed was a member of the Hilton Head Natural Resources Advisory Group and served on the South Carolina Heritage Trust. His hand could be seen in the new natural design of Hilton Head Island and in the enhanced sea turtle protection there. He also helped improve the signage, visitor displays, shelters, and parking areas in the refuges of the Savannah Coastal Wildlife Refuge system. He did the same for the Pinckney Island National Wildlife Refuge adjacent to Hilton Head Island. One person *can* make a difference. He chalked up his success to his three core beliefs: (1) treat others as you want to be treated, (2) be kind to everything that lives, and (3) do all things with love. Ed died in 2009, leaving many legacies, one of which is a much more environmentally friendly Hilton Head Island.

Two Nations

The United States has a split personality when it comes to sea turtle protection. The government protects sea turtles, especially on the nesting beaches, but at the same time it promotes commercial fishing and puts sea turtle protection secondary to profits from the fishing industry. Costa Rica is another country with a split personality. It is known throughout the world as a leader in biodiversity and conservation. It formed a national park system in 1970, and 25 percent of its land is in parks and other protected areas. However, it also fails to protect sea turtles in the ocean, does not enforce its TED regulation, allows illegal fisheries to operate, and provides a classic example of what happens when overdevelopment attacks a sea turtle nesting beach.

Tamarindo, aka "Sodom," as in the biblical tale of Sodom and Gomorrah, is a small town situated in the middle of Parque Nacional Marino Las Baulas in Guanacaste Province on the Pacific coast of Costa Rica. It once hosted many nesting leatherback turtles, but no longer. In 1990 Tamarindo was described in one of the few guidebooks available in the pre–tourist rush days, *The New Key to Costa Rica,* as "a wide, white-sand beach with a large estuary—a favorite with surfers and windsurfers.

The estuary has been made into a wildlife refuge. Leatherback turtles nest on the beach, August to February. Many baby turtles have been crushed as they scramble to the ocean by inadvertent tourists because it is hard to distinguish them in the dry, loose sand high up on the beach."

That problem does not exist anymore because leatherbacks nest elsewhere. Now leatherbacks nest on Playa Ventanas and Playa Grande to the north of Tamarindo and Playa Langosta to the south. They do not come ashore on Playa Tamarindo for several reasons. First, there is little sand left on the beach. Builders took it away to make cement for the many hotels that fill the town. Second, plenty of pollution is dumped directly into the ocean from those hotels as well as from all of the stores and condominiums in town. There is no central sewage treatment system and no individual systems for the hotels. The waters along the beach have very high fecal coliform bacteria counts, and there is plenty of chemical pollution to go around as well. Any self-respecting turtle coming back to its natal beach on Tamarindo would react negatively to the new smells in the water and go somewhere else. Third, the town is lit up at night like a prison. Crime is a big deal in Tamarindo, so every hotel, restaurant, store, and house lights up its property to foil thieves and, at the same time, attract customers.

Pamela Plotkin of Cornell University has conducted a series of campaigns sponsored by the Leatherback Trust—our nongovernmental organization (NGO) that is trying to save the turtles—to turn out the lights in Tamarindo. Some people have cooperated, some of the time. But the positive effects of the campaigns wear off after a while, and people forget to turn off their lights or to paint them red. So you can see Tamarindo from 10 miles out to sea, and when you are walking on Playa Grande you get blinded by the lights spreading from the seashore to the top of the high-rises ringing the hills above the town.

In the early morning the hotels hire young children to walk the beach to pick up dead hatchlings that have been attracted from Playa Grande to Tamarindo by the lights. The hatchlings leave the beach and swim to the left toward the lights instead of out to sea. They get washed onto the Tamarindo beach and go around in circles until they are exhausted. The hotel owners don't want the tourists to get upset, so the children fill up their buckets and take the hatchlings away before the visitors come down to the beach. Frank Paladino, my former student and now colleague, watched them do it.

I first came to Tamarindo in 1990 with Randall Arauz, then a student at the University of Costa Rica. At that time there were only four hotels and a couple of places to eat; Cabinas and Restaurant Zully Mar at the center of town was our favorite. It had a great open-air restaurant behind the beach with a lovely view of the ocean and terrific wooden doors on the bathrooms carved with lovely ladies in scanty bathing attire. The *cabinas* (small hotels) were across the street and were simple but clean, with more doors showing Costa Rican scenes. It was Jorge, the owner, who told us about the leatherbacks nesting on Tamarindo beach. He collected their eggs for several years to get enough money to buy land and construct his *cabinas* and restaurant. Unfortunately, the restaurant was within the 50-m (164-feet) public zone and after he sold it, the new owner had to tear it down and rebuild back 25 m (82 feet) in what used to be the parking lot. It lacks the charm of the original, the food is not as good, and the prices have risen quite noticeably. Somewhere in that process the bathroom doors disappeared. Too bad, as I hoped to get them for our biology station on Playa Grande.

One of the other hotels was the Hotel Dolly, which was down the road close to the beginning of the little village and just off the beach. It was even simpler than the Zully Mar and hosted a lot of mosquitoes. Leatherbacks used to come up and nest right under the buildings. People took their eggs there, too, as they did throughout the beach. The best place to stay was the very pretty and pleasant Hotel Diriá in the center of the village. The Tico who owned it liked the turtles. The Diriá had air-conditioned rooms and a lovely pool and garden. It even had a telephone by the front desk that you could use to call out. We could sit by the pool and watch the green iguanas and ctenosaurs (spiny-tailed iguanas) climb over the rock wall and up into the trees. It was another great place to watch the sunset. This was the preeminent Tico resort. In fact, Tamarindo was almost exclusively a Tico vacation spot.

All that has changed. U.S. and European tourists have discovered Tamarindo. The only thing that hasn't changed in Tamarindo is the Hotel Dolly. The Zully Mar built a big multistory hotel and changed the old *cabinas* into a strip of shops and a massage center. The doors are gone there, too. German investors bought the Diriá, renovated it, and sold it to a multinational corporation that did another "upgrade." The main street is now completely lined with hotels and shops for the many foreign tourists who clog the street with cars and the sidewalks with bodies. The Lonely Planet guidebook calls the town "Tamagringo" and does not have much nice to say about it. Now you can get anything in Tamarindo, and I mean *anything*—beer and rum as before, drugs of all kinds offered on the sidewalk, t-shirts and tourist fare, handmade jewelry, a male or female friend for the night (or the hour), pretty much anything except a nesting turtle.

Of course, there is plenty of crime to go around. If you have a rental car, you need to be sure to pay the fellow who is standing around pretending to guard the parking area, or you will have a broken driver's-side window, at minimum. Watch your wallet, because there are plenty of pickpockets. Be careful crossing the street, and don't swim in the ocean unless you are immune to bacterial infections in your gut. Lovely place, Tamarindo; who needed the turtles anyway? The funny thing is that it could all have been predicted. Leatherback turtles used to nest on Playa Flamingo, too, but they left that area when it was developed before Tamarindo.

I used to think that Tamarindo was about as bad as it could get, but that was before I went to Laganas in Greece. If Tamarindo is Sodom, then Laganas is Gomorrah. It is like Tamarindo, only several times worse. Laganas is a beach town on the island of Zakynthos. Greece is another country with a split personality about sea turtles. I praised it in my book *Sea Turtles* because of the great work being done for sea turtles by two NGOs there, ARCHELON, the Sea Turtle Protection Society of Greece, and MEDASSET, the Mediterranean Association to Save Sea Turtles. However, the government has lagged far behind in protecting the turtles. It has allowed extensive development right on loggerhead nesting beaches on Crete and does little enforcement of regulations for sea turtles in the Mediterranean Sea.

Zakynthos hosts the largest population of nesting loggerheads in the Mediterranean Sea. Loggerheads used to nest on six beaches: Laganas, Kefalonia, Sekania, Daphne, Gerakis, and the island beach of Merothanisia in Laganas Bay. Over the last several years, Laganas has boomed as a resort destination, especially for young English men and women. As a Wild West town it puts Tamarindo to shame. Again, anything goes in Laganas, from strip joints to loud and bright bars. The English youths let go of all their inhibitions, drink themselves into oblivion, and things get wild. The sand on the beach is compacted from too many people and vehicles, and loggerheads seldom go there to nest anymore.

In 1984, a presidential decree promised to protect the nesting areas on Zakynthos, and in 1999 a new law established a National Marine Park in the area, including the beaches and the bay. That did not prevent the disruption of Laganas beach, which turtles have essentially abandoned, and the illegal development of Daphne beach, where homeowners make it very difficult for sea turtle teams to work. Only after the European Union stepped in after 2005 did Greece finally adequately fund the Marine Park. Hopefully, protection will now be provided for the five remaining beaches. Laganas is lost, but there is still a lot of nesting to protect on the other beaches.

Lessons Learned

When I was serving with the secretary of the Army I was impressed with a process that the Army (as well as other federal agencies) uses, called lessons learned. After every practice battle at Fort Irwin in California the enlisted soldiers and officers would gather together to determine what lessons had been learned from that exercise. In terms of what we know now about nesting beaches we can list several sea turtle lessons learned.

1. Keep the beaches natural. If a sea turtle nesting beach has not yet been developed, leave it alone. Put it into a park or a reserve of some kind and regulate use of it.
2. If a turtle nesting beach already has houses or hotels on it, stop development now. It may be necessary to grandfather-in houses that are already there and have owners participate in the protection of turtles and their nests.
3. Turn off the lights. There should be no lights shining on a turtle beach. That should be pretty clear from this book. If there are houses or hotels behind a beach, be sure that they keep their lights off. As seen on Jekyll and Hilton Head islands, local lighting ordinances can be effective in keeping the beach dark for the turtles.

4. Stop the pollution. Surface and groundwater pollution in and around turtle nesting beaches will eventually be the death knell for the turtles using those beaches.
5. Stop armoring beaches. Sea walls, rock breakwalls, rock rip rap, geotextile tubes, and similar devices make beaches inaccessible to sea turtles. Face it, seas are rising (see chapter 12) and beaches are moving landward. Let them move and let sea turtles nest on them.
6. Collect the data. It is very important to keep track of the turtles and their nests to determine the long-term trends in the nesting population. Turtles are long-lived animals and their population trends have long lag times. It is necessary to track their reproductive success and population sizes for many years to determine if conservation measures are reaching their goals.
7. Never give up. One good thing about sea turtles is that there are probably large numbers of juveniles in the ocean pipeline. So even if things look bleak on the nesting beach, keep up the effort, and in a few years hopefully you will see the fruit of your efforts.

9

NESTING: TAKING BACK THE NIGHT

As the sun sets over the palm trees lining the lagoon, it sends yellow and orange rays through the rain clouds over the tropical forest behind the beach at Tortuguero. The nearby surf crashes ashore as fifty to a hundred green turtles wait just beyond the breakers for darkness to come. A red sky then envelops Cerro Tortuguero, the mountain at the north end of the beach. A black shadow is cast by that magnetic mountain, a dead volcano. That shadow foretells the black night to come. Black sky, rain clouds broken with lightning, torrential downpours, and black sand make for a mysterious and sometimes terrifying walk down the beach. But this is the magic place, the place where the green turtle survived and made its last stand in the Caribbean Sea.

When Archie Carr first found the beach, there were hundreds of turtles nesting each year. The local people ate turtles and their eggs, and people from the port city of Limón, about 50 km (31.3 miles) south, came up in the open boats they call *cayugas* to collect turtles for the slaughterhouse. Green turtles had disappeared from many beaches in the region, and things were looking bad for them at Tortuguero. It was 1950.

In 1955 Archie set about organizing one of the longest-running studies of the population biology of any animal. Larry

Ogren, David Ehrenfeld, Harold Hirth, and a continuing flow of graduate students and volunteers counted turtles, incubated eggs in hatcheries, transplanted hatchlings to other beaches, and, most importantly, established a conservation ethic in the local village. By involving villagers in the project, Archie injected money into their economy and raised the stature of the turtles. Dramatic fluctuations marked the population for many years. However, after three decades of protection and research the numbers started to climb. By 1990 there were 7,000 or 8,000 turtles nesting in a season. By 2008 there were 20,000.

Tortuguero is perhaps the most famous sea turtle beach in the world. Walking down it at night you cannot see your feet in front of you. You walk by feel because the black sand blends into the black night. I was patrolling the beach with Molly Lutcavage, Frank Paladino, and Mike O'Connor a few years ago, and as we went along the first person in line disappeared, just plain disappeared. We heard a thud and a small cry from below. The edge of the berm had been found. It had crumbled away and another biologist was in the waves. With a little help from friends we were all back on our way.

The Nesting Process

When a green turtle comes ashore at Tortuguero, she sticks her nose into the wet sand and then continues to move forward until she finds the "right spot." No one knows what the cues are for that right spot. Is it the wetness of the sand, the temperature, or some smell? There are lots of suggestions but no strong data that yet answer that question. Sometimes the female turtle does not find the right spot or gets scared away by lights or noise. Sometimes unwary tourists use a big, white light to see where they are going and shine it on the turtle. Then she turns around and heads back to the ocean. She leaves a half-moon trail on the beach that can be seen and counted the next morning. She will hopefully try again somewhere else.

When we walk on the beach at night, we walk in total darkness, but when we need light we use red lights. They don't bother the turtles as far as we can tell. Many sea turtle projects are supported by a program called Earthwatch, in which everyday people volunteer to help patrol the beach and count the turtles and their eggs. Volunteers on the beach are great because they help to keep poachers away and help the researchers collect the data needed to keep track of the turtles. In addition, contributions from the volunteers help pay for the expedition. It is a good situation for the volunteers, the scientists, and the turtles. These volunteers learn quickly how to navigate without lights.

It is interesting to take people out on the beach for the first time without a light. People regularly start to freak out. But after a while the volunteers love being able to stroll under the stars and see where they are going without a flashlight. Of course, everyone falls over a log now and then or into the abandoned body pit left by a nesting turtle, but eventually you learn to avoid most of these hazards.

The inevitable question from the volunteer is "How are we going to see the turtles without a light?" We always answer, "Don't worry, you can't miss them." Sure enough, a few hundred meters down the beach we come across a trail coming up out of the surf that looks like it was made by a tractor or tank. The turtle has stirred up the sand, and the contrast between the wet sand that she has turned up with her flippers and the drier sand on the surface cannot be missed, even in the dark. So we creep up the trail to the top of the berm, and there she is, as big as a small car, or so it seems. "Wow, she is so big, you really can see her without a light!" remarks the volunteer. The human eye is a wonderful thing. We really can see in the dark, and the beach and the turtles are so much more interesting that way.

As you walk along that beach sometimes you see moonlight gleam off the wet carapace of a green turtle as it emerges from the surf. From a safe distance you watch as it crawls up the beach and looks for a place to dig. A green turtle starts her nest by throwing sand with her front flippers. She is digging a body pit. She scoops large piles of sand and throws them behind her. She keeps that up for 30 minutes or more until she has lowered herself about 20 cm (8 inches) into the sand so that the top of her shell, the carapace, is even with the beach. The effect is that the beach is dotted with what appear to be artillery shell holes or shallow foxholes. Even with good night vision you can trip into the holes left by the turtles. I once fell into so many nest holes that my knees gave out. That's a major reason why the edge of the berm is a better place to walk: you may fall harder, but you fall less often.

After the green turtle finishes the body pit, she starts to dig with her hind flippers. The rear flippers are like big baseball catcher's mitts. First the right flipper and then the left does the work. Two graduate students, Annette Sieg and Eugenia Zandona, discovered something fascinating about this process: leatherback turtles may be right- or left-flippered just as humans are either right- or left-handed. In her previous research, Eugenia discovered that howler monkeys tended to be right-handed. She and Annette decided to determine if leatherback turtles had a flipper preference in digging. They and their assistants recorded the flipper with which the turtle started to dig. After compiling data on multiple nestings of several hundred turtles, they were able to determine that there was a slight tendency for more turtles to be right-flippered. That was true both in the Pacific at Playa Grande and in the Caribbean at Pacuare beach.

It can be mesmerizing to watch the turtles dig. The flipper comes over the centerline and scoops up some sand. Bracing with the left flipper the turtle brings the right flipper back and flips the sand away. Then she places that flipper over the loose sand and compacts it while bracing for the left flipper to do its work. That goes on for 30 minutes or so until a flask-shaped nest chamber is completed. It looks like the turtle is watching the entire process, because it seems so precise. However, the behavior is really hardwired in her brain. The turtle can't turn around to see what she is doing. If a turtle comes ashore missing one of the flippers, owing to a shark bite, she will go through the same process with the stump of the flipper doing the same motion without picking up any sand. Of course, this means that the process takes twice as long. If both flippers are bitten off, she will go through the process and eventually drop her eggs on the sand because there is no nest hole. Sympathetic biologists often dig the hole for her. My assistants and I have done this for an old friend, "Stumpy." It saddens me that I have not seen Stumpy for 10 years. I wonder what happened to her. Did she move elsewhere? Did a shark finish her off? Was she yet another victim of too much commercial fishing?

Once the eggs are laid, a turtle covers the nest in the reverse process, which also appears careful and laborious, as if she has an eye in the back of her shell and is watching the sand fill in the nest cavity. First, the right flipper picks up a scoop of sand, swings it over the throat of the nest hole and drops it. The turtle pats down the sand with the flipper and retracts it. Then the left flipper does the same. After the nest cavity is filled, she uses the back flippers to compact the sand. She packs it in with her flippers while periodically testing for firmness by sticking her tail into the sand. When the firmness is right she stops, waits a moment or two, and switches to throwing sand with her front flippers.

Huge amounts of sand fly through the air and start to cover the area. This is one of the fun times to have a new student or volunteer. Sometimes we let them measure the carapace length or width of the turtle and "forget" to warn them about the covering process. There is no better initiation to sea turtle biology than to bend over a turtle when she suddenly starts throwing sand with those big front flippers. The sand can go twice the length of the turtle. A couple of flipperfuls of sand sends the person jumping and gets the rest of the team laughing.

There is a great video of me measuring a leatherback, in which she reaches over with her right front flipper when she starts to cover and catches my leg. All you see is a blur as you hear a shout and I end up 5 feet (almost 2 m) behind the turtle, a little sandy but otherwise fine. It was the students' turn to laugh at me.

The nesting process is pretty much the same for all sea turtles. Ridleys are too small to cover their nests with the hind flippers alone, so they do the "ridley dance." They are so small they have to get their whole body into the process and actually lift off the ground as they push down with the rear flipper. In fact you can hear a ridley covering from quite a ways away. A great thump-thump sound comes from the beach. When you walk up you see the turtle doing its little dance. It actually gets up on its flippers and bounces its body down from side to side on top of the nest. Being small, it does not have the strength to compact the nest in any other way.

A female leatherback seems be on autopilot when she covers her nest. But there is a hint that her brain is fully engaged. Periodically she sticks her tail into the sand as if to test the compactness. Then she continues her efforts, adjusting according to how far down the sand has compressed. You can tell when she is finishing, because she uses both flippers at once to press the sand and complete the job.

The Problem of Poaching

Once the nest is closed and compacted the turtle starts the wild "throwing of the sand." As if rowing a boat, she uses both front "oars" to throw sand around, covering up the nest. By moving a bit closer to the nest after every few throws, she maneuvers around the beach, making a large disturbed area. A green turtle leaves a pit with sand thrown around it, but a leatherback leaves an area 5–10 m (16–33 feet) in diameter, with the sand all mixed up and two or three places that look like the location of the nest itself. An hour or more later even a trained person has great difficulty finding the exact location of the leatherback nest.

Of course, the covering process does not always work since raccoons and dogs can smell the fresh scent of many nests. Poachers take a long stick that is sharpened at one end, poke it into the loose sand, and feel for the "throat" of the egg chamber. Then they dig up the eggs and put them into a sack. Poachers have used this technique effectively for more than five decades.

I have talked to poachers over the years and have heard some interesting tales. In June 1980, Stephen Morreale and I were on the beach at Tortuguero and we met a tall, muscular guy walking along with a canvas sack full of eggs over his shoulder, a long stick in his hand to probe for sea turtle nests, and a really big machete in the other hand. He told us he was just taking a walk down the beach. We told him we were researchers and asked him not to disturb our markers and the thermocouple wires that we used to measure sand temperature. We didn't tell him that the wires led right to a clutch of eggs below the surface of the beach. Later we found out that his father was in jail for killing a couple of people on the beach a few years before and that the son was supposed to be a lot meaner. If we ever met him again, our friends said, we should keep our distance and not talk to him. They noted how brave we were. Well, it's easy to be "brave" when you don't know you are talking to the bad-natured son of a murderer.

Every poacher is a real problem. But poachers do not work alone. It takes organization, logistics, middlemen, transportation, and developed markets to handle the millions of eggs that are removed from beaches every year. Frank Paladino and I had direct experience with that level of effort on Playa Langosta. The year was 1990 and Frank had to negotiate with the poachers, who covered the entire beach, so that we could conduct leatherback studies on only one 100-meter section of the beach (about 330 feet). I first met Frank over 30 years ago in the hallway of the science building at Buffalo State College when I was an assistant professor. He came by looking for the Department of Biology of the State University of New York at Buffalo so he could apply for graduate school. Frank was off to a bad start. He was at the wrong university, but I knew Buffalo State (formally known as the State University of New York *College* at Buffalo) was often confused with SUNY at Buffalo, which lies about 5 miles away. I was mildly amused, and as we had some friendly competition with folks across town I questioned Frank about his goals. Upon hearing about his interests I remember thinking, "He should be going here." The rest is history. Frank went on to get a master's degree with us, a PhD at Washington State, and ultimately became department chair of biology at Indiana-Purdue University at Fort Wayne.

Back to 1990 and the beach on Playa Langosta, Frank and I continued our search for leatherbacks to study. Frank, by that time a young professor, was beginning his now-famous studies of leatherback metabolic rates. With a net to catch a turtle, a huge tripod, a winch, and a scale, we looked laughable to the poachers. To add to the odd scene, a television crew from WHYY in Philadelphia came along to film the process. That group was making a series of documentary films on dinosaurs. They wanted to get the leatherback story because Peter Dodson, the world's leading dinosaur biologist and a professor at the University of Pennsylvania, had told them that our research showed that leatherback turtles were a good model for the physiology of big dinosaurs.

The beach was thick with poachers. It turned out that the foreman of the farm "Finca Pinilla"—where we were staying—was the leader of the poaching ring. By day he was a gracious host and set us up in the bunkhouse, as directed by his manager. At night he rode the beach on his horse and made sure that each of the poachers, who had come from the local towns of Villa Real and Pinilla, stayed in their 100-m sections. As long as they did, they were allowed to take all of the eggs deposited there.

The foreman had a friend who would come in a truck and take the eggs once they were collected. This middleman then drove the eggs to Santa Cruz, Liberia, or San José to sell to bars and supermarkets. In supermarkets you could find the large leatherback eggs sitting on a little tray wrapped in plastic and labeled as coming from Ostional, where the legal harvest of the much smaller olive ridley eggs takes place. The farm foreman, now the poaching leader, got a percentage of what each individual poacher got for his sacks of eggs. It was a very organized system. With this system in place, every clutch of eggs was taken from the beach. In fact we only saw one hatchling—not one nest, one hatchling—come off that beach that season.

When we first arrived on the beach, with film crew in tow, the poachers were nervous and edgy. They were breaking the law, but it had never been enforced. Now Americans with a film crew had shown up. Frank said that he would talk to them.

He went off into the darkness and soon came back with an arrangement: we would not film them, nor interfere with their poaching, and they would let us work on our turtles—after the poachers took the eggs. In addition, Frank would pay the "owner" of that beach section 100 *colones* (Costa Rican currency, about eight to a U.S. dollar). It was a deal with the devil, but it was the deal we made. Every night for a week we weighed turtles and collected their breath (for analysis of their respiration). You can see us in the program *Flesh on the Bones,* which still airs periodically.

After that nesting season ended we spoke to the owner of the ranch that included the beach, trying to encourage him to stop the poaching for the next year. It turned out he was not a big fan of having all these people on his beach. However, he said he had once driven them off and the result was pretty bad. They burned his pastures and he had no forage for his cattle during the dry season.

We came up with a plan, one that would keep him a bit removed from the situation. He would make a donation to the turtle fund at Drexel University. Then we would hire a policeman with a big, silver handgun and a gold badge the next season. Our cop chased away all of the poachers so that we were able to do our research without interruption. More importantly, we were able to start a population-monitoring program on the beach with the World Wildlife Fund. That program continues to this day, and the beach is now part of a national park. Leatherbacks lay their eggs in peace, and now a whole lot more than one hatchling emerge from the sand and run to the water.

A Truck Full of Eggs

It is illegal to take eggs from sea turtles in Mexico, and yet sea turtle eggs are a big business in that country. Some of the poaching is run by local drug cartels, so it is also a dangerous business. A few years ago I read a small article in the *Philadelphia Inquirer* about how wildlife authorities in Oaxaca, Mexico, had confiscated a semitruck full of sea turtle eggs. My colleague Georgita Ruiz was from Oaxaca, and I wondered if she had a hand in that seizure. Georgita had worked with me in Buffalo during early studies on temperature-dependent sex determination. Trained as a veterinarian, she worked at Playa Nancite, along with Merry Camhi, to collect field data on sex determination in olive ridley sea turtles. She carried out the lab portion of that work in Buffalo.

Georgita became the federal *diputado* in Oaxaca. The *diputado* is the federal wildlife director for a state (in Mexico, the federal government plays a more powerful role in some areas of governance). Georgita was the first woman to hold such a post, and she had won the support of a team of tough male wildlife officers. She set about to actually enforce the wildlife laws there, something many still see as bold and courageous. One night her guys were working a roadblock on a tip that eggs would be heading to market. Along came a big semitruck escorted by a police car. Georgita's crew stopped both the truck and the police car and said they needed to see what was in the truck. The policeman threatened to arrest them and would not let them inspect it. Georgita knew things could get dicey, so she had reinforcements close by: the Mexican Army. They were immediately called in, and the policeman quickly drove away. At the approach of the soldiers the men driving the truck took off into the woods and got away. How many eggs does one truck hold? Georgita's men counted 526,000 olive ridley eggs!

While far from perfect, Mexico does not often get the credit it deserves for its work on behalf of biodiversity. Mexican Marines patrol leatherback and olive ridley nesting beaches and have actually shot poachers who were stealing eggs. Since the poachers are often armed with machetes and sometimes guns, we should not be surprised that beaches in Mexico can be dangerous places. On one beach poachers came into camp, tied up the biologists, beat them, and took their computers and 4×4s. In Oaxaca, one of Georgita's inspectors was shot in a bar by a poacher. On Rancho Nuevo on the Gulf of Mexico, poaching of Kemp's ridley eggs was finally stopped only when the Mexican Marines were deployed to stop the poachers and guard the biologists and their camp.

The Tourist Effect

It is a common fact of life on a turtle beach that when you get rid of poachers they often are replaced by tourists. Tourism can be a mixed blessing and has matured quite a bit in many places. In 1989 at Playa Grande, tourists took pictures of leatherbacks at night with flashes popping. Guides actually encouraged tourists to ride the turtles. When Las Baulas Park was formed by presidential decree and Playa Grande was included in it, tourism was brought under control. María Teresa Koberg, the force behind formation of the park and the first park director, began a twofold program: she patrolled the beach to stop poaching and she worked with the local communities of Tamarindo and Matapalo to train former poachers as ecotourism guides. Now there is a highly regulated form of tourism on the beach. The guides associations of Matapalo and Tamarindo work with the park rangers and biologists to locate the turtles, protect them, and take groups of tourists out to visit them. The basic group is composed of thirty people who walk single file in the dark behind a guide who takes them along the water out to where the turtle is nesting. The guides split the group into two and take fifteen people at a time up to see the turtle. The guide uses a red light as people are brought up in silence behind the turtle. People cannot go on the beach alone and cannot approach the turtle before she is done digging her nest or after she finishes laying her eggs and covering them up with her hind flippers. No one goes in front of the turtle's hindquarters. Each association is allowed two groups, so only 120 people a night get to see the turtles.

While the turtle is dropping her eggs and in her "nesting trance," a biologist checks for identification tags and measures the turtle's carapace length and width. Another biologist moves some sand, making an open hole. Through this hole the eggs can be seen coming out, allowing the biologist to count them. The guide then brings the tourists up close to see the eggs drop, looking through the hole made by the biologist. Once the turtle starts to cover the nest, the guide fixes the hole and the tourists are taken back off the beach.

This ecotourism project generates considerable income for the local cooperatives. Each person pays $20, which includes the $10 entrance fee to the park. There are as many as 120 tourists on the beach each night of the season from October 15 to February 15. That is about $100,000 in income for each cooperative over the season. In the Matapalo cooperative a third of the nightly revenue goes directly to the guides who work that night, a third to the cooperative to be shared among all the guides, and a third into a fund for community development. Each night some of the guides walk the beach looking for turtles and some take out the tourists. So a turtle guide makes a nice little income while wearing a clean shirt and taking a walk on the beach with some tourists. He or she does not have to get down in the wet sand and poach. Gone are the days of living like a coati or raccoon.

Instead, they are proud conservationists. The guides and their families are now the strongest proponents of the park.

There were a lot of growing pains in the guide system at Playa Grande. The guides did not have very good English skills, while most of the tourists spoke English. My crew helped by holding English courses for them. The guides did not know the biology of the turtles, so Frank Paladino provided turtle talks to educate the guides about the turtles and the nesting process. After a few years most of the guides became better informed. Now our turtle team gives the guides updated information each year.

Some people argue that the presence of tourists, however controlled, affects the turtles. We carried out an extensive study in which we timed the behavior of the turtles with and without tourists present and found there was no difference in the tendency of turtles to return to the water without nesting, in the timing of different behaviors, in the egg-laying process, or in the hatching success of the clutch. As long as the groups of tourists were managed correctly and were under the control of the guides, the turtles carried out their nesting process as if no one was there.

Similar programs have been instituted with great success in other areas. In South Africa, for example, turtle tours have become very successful and lucrative. The tour operators help to collect data on the turtles, and the tourists get to observe the research activities as well as the turtles. Bonus benefits include voluntary contributions from tourists that help pay salaries of local staff, purchase equipment such as tags and satellite transmitters, and support educational programs in local schools.

Having ecotourists underfoot on the beach is certainly better than having poachers there. In addition, as long as a turtle tour is conducted in a safe and ecologically responsible manner it can benefit the turtles. Ecotourism supports local communities, builds a support base for the turtles and their protection, and gives sea turtles celebrity status. A high level of public awareness and empathy are needed if sea turtles are to be preserved in this century. There is no stronger advocate for sea turtle conservation than a volunteer or a tourist who has had an up-close and personal experience with a sea turtle. Those people will be the base on which we can build a successful global program of sea turtle conservation.

10

LAS BAULAS: THE LAST HOPE FOR PACIFIC LEATHERBACKS

Las Baulas is a national park in Costa Rica with the formal name of Parque Nacional Marino Las Baulas. The word *baulas* is Spanish for "leatherback." Las Baulas is located on the Pacific coast of Guanacaste Province near the Costa Rican town of Tamarindo. The park consists of four main beaches, some mangrove estuaries, and a few hills. Leatherbacks nest on three of the beaches. The marine section of the park, Bahía de Tamarindo (Tamarindo Bay), stretches 12 miles out to sea. The park hosts 90 percent of leatherbacks on the Pacific side of Costa Rica. These three little beaches have as many nesting leatherbacks as all the Pacific beaches of Mexico. But the turtles here are not safe. Las Baulas is at the center of a maelstrom of controversy, and its future will define the future of conservation in Costa Rica in the twenty-first century.

Discovered as a major nesting zone in the 1980s, first decreed a park in 1991, and established by law in 1995, Las Baulas came close to being four beaches without turtles on several occasions. Now it is at the center of a Costa Rican cash grab, not too different from the gold rushes of old. The "precious metal" this time? Developing the coastline of Guanacaste.

Many parts of the Caribbean coast of Costa Rica are wet, rainy, and often unpleasant. Not ideal conditions for building a resort. The Guanacaste coast, on the other hand, is dry and sunny 6 months of the year. During the other 6 months—the rainy season—there is plenty of sun most of the day. Add to this Guanacaste's white sand beaches, which curve between headlands of striking proportions.

Making a Park

María Teresa Koberg, a Costa Rican woman who loved nature, decided to save the turtles of Las Baulas. She is the real hero of this tale. There is a lot of controversy about who established the park and who was involved in the early efforts to save the leatherbacks. It is amusing to hear some of those stories, which are sheer fiction. Beyond María Teresa, the people who deserve the most credit are Mario Boza, Clara Padilla, Peter Pritchard, and Frank Paladino.

Prior to the 1990s Playa Grande was like a scene out of the Wild West. There were no rangers, the beach was uncontrolled, poaching was rampant, people rode the turtles and took flash pictures, and houses were built close to the beach. Even after being established, Las Baulas was what conservationists call a "paper park." Nothing was actually being done to protect the area. If you post a speed limit but everyone knows that there are no police, people speed. If you say an area is a park and poachers want the eggs and there are no guards, they take the eggs.

When María Teresa began her quest, plans existed for a major development of 300 houses and a luxury resort for 5,000 people, with a casino, nightclub, hotel, condominiums, and a yacht club along the beaches that became the Las Baulas. She started by walking the beach at night to talk with the poachers. The poachers had machete fights over the eggs and were often drunk. María Teresa was a missionary for conservation. She converted Doña Esperanza Rodríguez, the matriarch of a family that was long the major source of poaching on Playa Grande. She paid Doña Esperanza to keep poachers off the beach and to make daily nest counts on Playa Grande. In later years I checked Doña Esperanza's nest counts by simultaneously using scientific methods. Her data were within 5 percent of the "real" numbers.

María Teresa obtained national media attention for her efforts and promoted the idea of a park that would protect the area. She was opposed by many forces, including a local hotel operator who now bills his lodgings as "ecofriendly." María Teresa brought in Boy Scouts from both Costa Rica and Minnesota to patrol the beach and become advocates for conservation. The presence of the Boy Scouts embarrassed the poachers, and some stopped. Poachers know that what they are doing is wrong, and some go away when other people are on the beach.

In 1990 famed biologist Peter Pritchard and members of the local community wrote a report that recommended formation of a park whose main purpose would be the protection of leatherback nesting beaches. Now it was more than one dedicated woman—a famous biologist was pushing as well. In 1990 Rafael Ángel Calderón won the national election and took office as president of Costa Rica. He appointed Mario Boza as vice minister of the environment. In one of the shining moments in the history of "good things happen to good people," María Teresa was appointed director of the National Sea Turtle Program. On June 5, 1991, President Calderón issued a decree, drafted by María Teresa and Mario Boza, that protected the beaches as Las Baulas de Guanacaste National Marine Park (Executive Decree No. 20518). The land limit was 125 meters (410 feet) inland from the high tide, and the ocean limit was 3 miles.

María Teresa became the first director of the park and established a protection program involving leaders and members of the Playa Grande and Tamarindo communities. Because of her parallel responsibilities as program director in San José, she brought in a young biologist, Randall Arauz, to help out at the park. He began an education program in the nearby villages. In 1992, Randall became the second director of the park and raised the research and conservation program on Playa Grande to a higher level. He spent a lot of time in the local village of Matapalo working with poachers and trying to convince people who did not support the park that the economics of the area would be improved by its designation. Gradually residents began to accept the park, as they realized that their concerns were being heard. Soon jobs opened up for truly ecofriendly turtle guides.

In 1991 Frank Paladino was able to get the Earthwatch Institute involved, a great organization that involves volunteers in research projects in biology, geology, archeology, anthropology, and other areas. Earthwatch volunteers on Playa Grande were force multipliers: instead of one lone student on the beach, there were three people. During 1992–1993, Frank Paladino began lessons for the guides, in Spanish, about helping turtles and controlling ecotourists. In addition, we helped park rangers construct the first house for the park, which served as the headquarters and a home for the director and his family. I brought Al List, a professor from Drexel, down to conduct botanical surveys and produce guidebooks of the estuary and dry forest. Through these the guides could enrich the tourist experience. That year saw the signing of the first in a series of cooperative agreements between the Ministry of the Environment of Costa Rica and Drexel University. The first of several ministry employees involved in developing the park then came to Drexel for English language training. José Quirós became director of the park and greatly improved on our training courses for the local guides. All was looking so good. Then the earth shook.

In mid-1993 the current minister of the environment, Hernan

Bravo, resigned to run for the Assembly. The new minister was of a different breed entirely. He eventually forced Mario Boza to resign. Our main ally at the national level was gone. The national election was held in February 1994 and José María Figueres was elected as the new president. I learned some horrible news soon after the election: in the time between the election and the installation of the new government the minister planned to have the president issue a new decree changing the park into a private refuge. I was not naïve about politics, but still I couldn't believe it.

Mario and María Teresa were in a bad spot. I wrote a letter to the minister complaining about all of the enforcement problems in the park and objecting to the park's rumored elimination and designation as a private refuge. At the same time someone in Costa Rica gave a copy of the letter to the news media. I have no idea how that happened, but the effect was profound.

Immediately the major TV stations went to the minister's office and asked him what was happening. Was it true that he was going to get rid of the park? He was surprised, embarrassed, irritated, and confused. His staff shrugged their shoulders and said they didn't know anything about it. The minister denied the rumors and stated that the park designation would never be changed from public to private. He accused "that Spotila guy" of being an outside agitator and banned me from the parks. Not wanting to mess things up, when it looked like things were getting better, I stayed away from Costa Rica for a few months. In May, I was invited back by the new minister of the environment, René Castro, and was officially named an advisor for Las Baulas.

In 1994, I contacted President Bill Clinton about the park. I knew the president through my brother John, who was a good friend of his and an official in his administration. The president instructed the State Department to send word of his support for the park to officials in Costa Rica. In 1995 the Costa Rican Legislative Assembly passed a law making the park a permanent entity. President Figueres signed the law during a symposium in the National Theatre in San José. Frank, Mario, and I were dinner guests that night at the Presidential House, the official "White House" of Costa Rica. That night I felt as satisfied as I ever have, but it was not to last.

Saving the Park

Unfortunately, the law had a typo. It stipulated that the park ran 125 m (410 feet) from the high tide line "into the water" instead of 125 m from the high tide line "away from the water." Even though geographic coordinates in the law made the boundaries clear, that phrase started a whole series of problems. Officials did not feel they had the authority to enforce the 125-m line, so people started building houses inside what I felt was Las Baulas National Park. On the good news front, more federal funds came to the park and Earthwatch England helped us obtain a grant from the Guinness Corporation's "Water of Life" program. Finally, a large dormitory building could be built for park volunteers.

When Rotney Piedra became director of the park in 1999, he was the first park director actually trained as a sea turtle biologist. He increased controls on the beach, worked with local landowners to improve protection of the turtles, and controlled poaching and tourism. He worked tirelessly with local residents and the guides. He was less successful in dealing with the park's fundamental problems: boundaries, home construction, road enlargement, and continued use of the beach by homeowners as their own property.

While others kept trying to get the typo fixed, Frank Paladino and I went to Washington and sought help for the park from the large conservation organizations. "Let's buy land," we thought. "As much as possible, and protect the area around the park." We found sympathetic ears but closed wallets. We pleaded that the extinction of the last major nesting population of leatherback turtles in the eastern Pacific Ocean was at stake. Someone told me I did not see the big picture. Another person said I was not in his organization's 5-year plan. A representative from another organization said that it was no longer budgeting money to acquire land in Costa Rica. Yet another group said that it already had a representative in Costa Rica and that it did not give out money directly to conservation projects like ours. We were frustrated but determined. We went to the U.S. Agency for International Development (USAID), which had a large program in Costa Rica in the 1980s. Too bad, the representative said, USAID was reducing its role in Costa Rica because the Contra war in Nicaragua was over. If we had only come by in 1988 when they had lots of money, we would have found help. We were frustrated and ready to give up.

A New Nonprofit Organization

By 2001 it was clear to me that if we wanted to help the park we were going to have to become entrepreneurial. That year we established the Leatherback Trust as a nonprofit organization. Its mission: to save sea turtles from extinction. In the first year we received about $10,000 in donations and established a website. As time went on we were able to pay salaries for rangers, help to establish a Ladies Association in Matapalo, fund environmental education in local schools and communities, fund development of a master plan for the park, fund the land acquisition process for the Ministry of the Environment, and directly purchase land to prevent development near the park. A grant from the Costa Rican office of Intel, the semiconductor chip maker, allowed us to establish an environmental education program in fifteen local schools.

The Ladies Association story is a great one. By 2003 the Guides Association had been up and running for a while and local support of the park was strong. Then in 2004 the International Sea Turtle Society held its annual symposium in San José, Costa Rica. Clara Padilla was one of the key "in-country" organizers. Clara was a local, living near Las Baulas. For a time she worked for us, running the Leatherback Trust operations within Costa Rica. Clara organized a visit to the park for the delegates in order for them to attend the Sea Turtle Festival at the park. She even got the minister of the environment to officially host the event. Two local land speculators saw an opportunity. They hired some local men to disrupt the event. Paid $20, the men stood around with signs denouncing the park, the minister, and the Leatherback Trust. Someone in the crowd shouted disparagingly about the minister and then the chant started, in broken English: "Don't trust the Leatherback Trust."

Bryan Wallace, my graduate student, went to the hired men and asked them why they were against the park. The answer was that they were paid to hold the signs but they could not read them, and that they would lose their jobs if they did not participate. They seemed to have no idea what the chant in English meant. The wives and girlfriends of the paid protestors were upset. Their children had practiced their festival skits, dancing and singing, for a month, and the demonstration almost ruined the festival.

The ladies took action. They decided to form their own cooperative to earn money and to be more financially independent. The Leatherback Trust paid the legal fees for their incorporation, and the Ladies Association was born. Rod Mast, vice president of Conservation International and co-chairman of the Marine Turtle Specialist Group, asked his brother, who had a business in Costa Rica, if he could find a vehicle for the ladies. He donated a part of the cost of a van, and the Trust paid the upkeep costs for the first couple of years. Now the ladies sell food to the tourists. The real estate developers still threaten their men with unemployment, but the formation of the Ladies Association has brought the temperature down.

It became clear that one of the major areas where we could help the park was in the legal arena. In 2004 we requested a ruling from the attorney general of Costa Rica concerning the boundaries of the park in the 1995 law. We provided information on the park that included maps, latitude and longitude readings from the law that established its boundaries, and an argument that the "into the water" phrase was just a typographical error. The ruling was issued in 2005, and it gave the rule of law to the true boundaries of the park so that enforcement and land acquisition could begin.

Land Acquisition

The law allowed for the buying of land, but it did not provide the funds to do so. The Ministry of the Environment pleaded poverty, so I grabbed Frank and we headed back to Washington. Rod Mast at Conservation International (CI) took us in. He helped us work with George Shillinger, who was doing double duty as a graduate student at Stanford University and staff member at CI. George envisioned the Eastern Tropical Pacific Seascape (ETPS)—a conservation zone including Costa Rica, the Galápagos Islands, and the coasts of all the nations through Ecuador. A UN grant for direct purchase of land provided $289,000. Then we all put our hearts and souls into the effort, and an amazing $3 million came from anonymous donors (I wish I could praise them publicly here, but they know that their money was used well). The Gordon and Betty Moore Foundation generously provided $2 million.

Land acquisition is not always nice. Sometimes people were happy to sell. Other times they wanted too much money. Still other times they didn't want to sell, even though their land was needed for some greater good (as we saw it, of course, not they). Much of the land had already been purchased by developers and speculators, so mostly the government took land from businesses by legal process—and we supplied the money to pay the landowners.

President Óscar Arias came into office in 2006 pledging support for an initiative called "Peace with Nature," in which Costa Rica would become sustainable and protect its natural heritage. Mario was pleased because the president had personally promised to help the park. Then one of the developers told Frank Paladino that the park was now dead and that Frank and I would be arrested the next time we came to Costa Rica. We did not believe that threat, but at about the same time park advocate Clara Padilla, Frank, and I received a series of telephone death threats. Mine were from one of the developers, and I recorded him on tape: "Hey, Jim, the next time you come to Costa Rica I will kill you and your wife and your children."

We went on but hit a snag. The minister would not sign the land expropriation orders, even though everything was in place. We brought more and more political pressure forward and finally we cancelled our long-term cooperative agreement between the Leatherback Trust and the ministry. That action caused a big stir in the Costa Rican media. The additional political pressure eventually got both the minister and President Arias to sign several expropriation orders in the spring of 2008.

Our move created some serious political blowback. The president supported the introduction of four bills by various representatives in the Costa Rican Congress that would reduce or eliminate Las Baulas Park. A long fight ensued that involved all of the environmental and sea turtle conservation groups of Costa Rica. Frank and I visited several key congressmen to brief them on the importance of Las Baulas. Costa Ricans from all over the country signed petitions and took to the streets in marches, and the park elimination and reduction bills died in the Environmental Commission of the Costa Rican Congress.

SALA IV

One of the forward-looking aspects of the Costa Rican Constitution is a unique provision that entitles citizens to a healthy and ecologically balanced environment. In 2008 the chamber of the Costa Rican Supreme Court that deals with environmental issues, the fourth chamber or SALA IV, began ruling on several lawsuits related to Las Baulas that had been wending their ways through the courts. First, the SALA IV ruled that Las Baulas was a valid park, and it ordered the minister of the environment to initiate and complete all land expropriations for the park immediately. Next, it ruled that the attempt by the local municipality (county) of Santa Cruz to impose zoning regulations in and around the park was unconstitutional. Then it ruled to stop all construction and invalidate all building permits in the park. Later, the SALA IV ordered an immediate halt to all construction within 500 m (0.3 mile) of the park, its buffer zone in the management plan. The court also suspended all building permits and environmental permits in that area, ordered a new environmental study of the entire area by the Ministry of the Environment to be done in 6 months, and for the second time, ordered the president to complete all expropriations in the park immediately.

We awaited the response of President Arias, who was proving to be less than "at peace with nature." He met with the landowners and pledged that he would not do any more expropriations. He also officially submitted a bill to the Costa Rican Congress from the Presidential House that would eliminate the park and replace it with a mixed private-public wildlife refuge. Frank told me, "Now I know what they mean when they call someone a 'tireless advocate for conservation.'" A day of rest, our Sunday, had not yet come.

Into the Future

The battle lines are drawn on Las Baulas Park. It is essential to the leatherback turtles in the Pacific, and its future is inextricably linked to the future of all of the parks of Costa Rica. The battle over President Arias's latest bill that would eliminate the park was fast and furious. The uproar in Costa Rica was heard around the world, and thousands of people and many international conservation organizations joined in to protect the park. The fight reached a crescendo in the final week of the congressional session before a new government took office after the national election in February. On April 29, 2010, hundreds of Ticos came out to the National Assembly, along with all of the national media. Channel 7, the major TV station, ran a story the night before, and we and our conservation allies took out a major ad in the newspaper. In the face of this display and a poll showing that 89 percent of the voters opposed the bill, the Environmental Commission held one last hearing and decided there was not enough time to vote on it. Yes—it died in committee! Yet the fight goes on as new threats continue to surface. Like Sisyphus, dozens of us push for the park to be forever preserved only to see the threats from 1994 return. I suppose as long as there are tireless advocates for development there will need to be tireless advocates for conservation.

11

OSTIONAL: THE EGG-STAINED SANDS OF COSTA RICA

A single-engine Cessna 180 angled around the hill at Punta Guiones on the Nicoya Peninsula of Costa Rica and banked low to fly north along the beaches of Playa Guiones, Playa Nosara, and Playa Ostional. It was the morning of August 24, 1970, and J. D. Richard and D. A. Hughes of the University of Miami were conducting their first aerial survey of sea turtle nesting on Guanacaste beaches. They flew at a height of 25 to 50 m (80 to 162 feet) and a speed of 130 to 145 km/hour (80 to 90 miles/hour) recording all of the indications of sea turtle tracks and nests that they found. They also counted turtles in the sea as they could. It was about an hour into their morning flight, and so far they had not seen much. They found no good turtle beaches between Punta Cuchillo and Cabo Blanco at the base of the Nicoya Peninsula and basically no activity between there and Punta Guiones. Playa Guiones was empty, as was Playa Nosara just above the river mouth. Rio Nosara is the largest river in Guanacaste and flows out of a series of hills that still support a semihumid tropical forest and now a protected area, Diriá Forest National Wildlife Refuge.

When Playa Ostional came into view it had very few nests, but Richard and Hughes noticed a large number of turtles out in the ocean. They swung out to sea and made a series of passes back and forth along the beach. Thousands of olive ridley sea turtles were swimming in the ocean and floating, seemingly resting. They flew on north, not knowing that they had just discovered one of the largest nesting concentrations of olive ridleys in the world.

The next day John Hyslop, a U.S. Peace Corps volunteer working for the fledgling Costa Rica National Park Service, went to Playa Ostional and observed a few hundred olive ridleys nesting on the beach. Fascinating, but only the tip of the iceberg. A month later, on the evening of September 26th, Hyslop watched in amazement as ridleys started to come ashore. First, there were a few turtles, and then they kept coming and coming. Richard and Hughes took to the air the morning of the 27th. As they came around the point, they could see thousands of olive ridleys ahead in the water, and the beach at Ostional was littered with turtles and covered with tracks. In terms of sheer biomass, it was the greatest display of reproductive effort ever seen by science. An estimated 50,000 ridleys nested on 4 km (about 2.5 miles) of beach in just a few days.

A Long Walk for Sand

In August of 1983 Ed Standora, Linda Lennox, and I were in Guanacaste to collect sand from the beaches of Ostional and Playa Nancite to measure the presence of bacteria and fungi. From the early days when the mass nestings, or *arribadas,* were discovered there, scientists knew that hatchling success of olive ridley nests on those beaches was very low, as low as 2 percent on Playa Nancite. Researchers had hypothesized that bacterial contamination was the cause, and we wanted to see what bacteria were present and what effect they might be having. After riding down the escarpment in Santa Rosa on horses and hiking over what was referred to as the "mountain," really a big hill at

50 m (164 feet) high but straight up, we walked down into Playa Nancite, where we found Steve Cornelius working the beach.

We collected our sand samples and watched the sun sink into the Pacific. Steve cooked up a supper of beans, rice, a little meat, and plantains over an open fire and oriented us to the beach and the *arribada* phenomenon. We learned that there were not very many turtles coming ashore in that year's *arribadas,* and Steve thought that it might be owing to something called an El Niño out in the ocean. The effect of El Niño on sea turtle reproduction was a new concept for biologists at the time. In Australia researcher Colin Limpus had started to figure out how massive the effect could be. He found that the water temperature in the central Pacific predicted the number of green turtles nesting on Heron Island in Australia. Basically, the warm water in the central and south Pacific reduced productivity, and therefore sea turtles did not get enough to eat. Less food meant no eggs. Cold water meant more food, which meant more nesting turtles.

The next morning we set off about 8 a.m. to carry 50 pounds of sand over the hill and down to the trailhead at Playa Naranjo. The trail up and out of Nancite had a gentle rise, the sun was not too hot, and there was a nice breeze when we broke out of the trees and onto the grassy slopes near the top. By the time we got to the base of the hill, where the horses were supposed to meet us, we had used up about half of our water. We had arranged for horses for three people, figuring that we would put the sand samples in the saddlebags. The guide had brought horses for himself and two others. He said the sand was too heavy to put on the horse with a person, so Ed and I volunteered to walk. We put Linda on one horse and the sand samples on the other. Linda and the guide took off and soon left us far behind.

Off we went along the muddy trail, over the first river, up the hill, over the second river, and up the escarpment. It was getting to be midday, and we were getting pretty hot. Our water was almost gone, so we started zigzagging along the road from the shade of one tree to another. Finally, at the top of our climb the land became flat. But the trees were gone and the sun beat down. At noon we had made about another 5 km (3 miles) and noticed that we had stopped sweating. That was more comfortable at first, but I looked at Ed's face and saw it was flushed with blood. Not a good sign. We were in the early stages of heat prostration. Ed was wearing jeans, so his situation was worse. There was a big tree about 100 yards up the road, so we headed for it and sat down in the shade. Ed took off his pants to dump the heat while I tried not to chuckle. We caught the cool breeze and started to feel like we might be okay. We agreed we'd sit and wait out the sun. Or maybe Linda would send help.

In about 15 minutes I heard a noise like an engine and the shaking of metal fenders. Ed quickly pulled up his pants, and out of the grass came an old, red Chevy Suburban. It rolled up to us and Linda and Douglas Robinson stepped out. Doug had come to camp to meet up with us and take us to Ostional. When we didn't show up with Linda, he came looking for us. He even brought water! We piled into the Suburban and headed to park headquarters. After a 30-minute cold shower we brought our core body temperatures down to normal and we were ready to set off for Ostional.

It is about 100 km (62 miles) from Nancite to Ostional. We moved along at a good clip until we got past Tamarindo, where the roads turned to dirt. After a couple of delays to let some cattle go by—as they still have the right of way—and to duct tape the battery cable back on the battery terminal, we were at the first river.

Three Rivers to Cross

The road from Tamarindo is still gravel today, and there are few signs leading to Ostional. You follow your nose and eventually come to a small river, at least during the dry season. Stay to the right on the gravel bed and you should be able to make it across. It takes a four-wheel-drive vehicle to get through those roads in the rainy season, when *arribadas* are at their peak. After a few more kilometers, you come to the second river. It is a bit smaller but has higher banks and is deeper. Driving another 5 km (3.1 miles) you come to the biggest river, Rosario. It has a footbridge way above you on the left that suggests that it is the only way to cross when the rains are seriously coming down. It is smart to sit and wait a while to see if a local comes by to show you the way across. If someone has gone through a few minutes before, you can see where the water is muddy and go there. If the water is clean and low then you can see the bottom and plot out your crossing. If no one comes to show you the way, you make a good guess.

Finally, you go up one more hill, swing down to the right and then left, and you come to Punta Rayo and the beginning of the Ostional Wildlife Refuge. Doug Robinson's Tico students who were working at Ostional had a supper of beans, rice, and coffee ready for us in no time. The students were bright, enthusiastic, and excited that their professor was there. They practiced their English and we our Spanish as they filled us in on the status of the *arribadas*.

Ostional was not exactly a village. It was only five or six houses on one side of the street and a couple on the other. The houses were really little more than simple shacks with wide planks for sides and thatch on the roofs. The people there did not own the land on which they built their houses and had little incentive to keep them up. The gaps in the sides of the houses were large enough for small birds to come in, and mosquitoes took pleasure in visiting all night.

At Nancite there was a field station to stay in with screens on the windows, and mosquitoes had to find their way in through the cracks in the floors. But then Nancite had its spiders. They were so big and so fast in the field station that Ed and I slept out on the beach. We spread a sheet out on the sand, and the ocean breeze kept the mosquitoes away. The next morning we awoke to find the tracks of coyotes all around the sheet. I guess they decided we were too big to eat.

Ostional had a bad feel about it. The people were not very clean, an unusual sight in Costa Rica. They lay around with nothing to do and seemed pretty rough. Doug said that they often fought over the eggs on the beach and were uncooperative when anyone discussed how to organize protection of the eggs or management of the turtles. It did not matter that they were doing something illegal and were putting their houses on public land. Ostional was theirs, the eggs were theirs, and they were going to keep everyone else out. They beat up one of the field assistants, slashed the tires on the truck, and interfered with scientists trying to tag the turtles.

By 1983 there was a truce of sorts in Ostional. Doug Robinson's students from the University of Costa Rica worked on the beach, and the locals left them alone. In 1979 Doug had guards posted on the beach to control the egg harvest and prevent violence, so things had calmed considerably by the time we were there.

The next morning we were up early and went out to the beach and collected sand samples before the sun rose too high in the sky. We were quick learners. The *arribada* had occurred some time before, and there were still a few vultures and coatis around but a lot fewer of them than at Nancite. We came back, got our cold showers and had a breakfast of chicken eggs, beans, rice, and coffee. The only water was from a shallow open well, and we were a little suspicious of it. It was okay in boiled coffee, but when Ed filled his canteen with the well water the first night I looked at him sideways. He left his canteen in the Suburban, and late the next morning he pulled it out to take a drink. It had been sitting in the hot car for several hours, and he noticed that it smelled pretty bad. He offered it to me to taste, but I said that I wasn't that thirsty and would get a drink later.

The next day Ed discovered that there were some interesting bacteria or similar creatures in the water. By the end of October he had lost almost 20 pounds to the little friends in his intestines. I finally convinced him to see a tropical medicine specialist at the Roswell Park Institute in Buffalo. After a monthlong course of antibiotics he had pretty much recovered.

The Management Experiment

Doug Robinson was frustrated that he could not get any serious protection for the *arribada* beach at Ostional. Costa Rica had outlawed the taking of turtle eggs in 1966, but there was very little enforcement and people took all the eggs that they could wherever sea turtles nested outside protected parks. Even in the protected Tortuguero National Park it was sometimes difficult to stop people from taking green turtle eggs, and sometimes people snuck in and took entire adult turtles.

As time went on the situation at Ostional got worse. More people moved in. They built houses and took eggs. Doug and Steve Curtis came up with an idea to try to help the situation. They proposed that the taking of eggs at Ostional be made legal but that stringent conditions be imposed to control the harvest. His proposal led to a lively debate in Costa Rica, but it was turned down by the government. In 1983 the government established the Ostional National Wildlife Refuge and organized a committee to determine what to do about the harvest of eggs. In 1984 Peter Pritchard got involved in the discussions and solicited opinions from biologists around the world about a controlled harvest at Ostional. Several of them were favorable, but others were more cautious or downright hostile. Discussions continued, and all the while the taking of eggs continued unabated. Doug Robinson kept pushing for a controlled harvest.

The biological basis for the plan was simple. Each *arribada* lasted only a few days. The eggs laid on day one were often lost, because egg laying on the subsequent days resulted in the digging up of many of the eggs deposited on the first day. The busted and exposed eggs that got dug up also may have contributed to the high bacteria levels in the sand. So why not let people collect the eggs on day one, since they would be lost anyway? The fifty or sixty people in the village would have an income, the situation would be controlled, and the eggs from Ostional would flood the market for eggs and reduce the poaching of eggs around the country. It was to be a market-based solution to poaching. The real test would be the effect of the harvest on the turtles' reproductive success. Scientific studies would test this new approach to conservation.

The Early Years

In 1987 the Congress of Costa Rica approved a new law to legalize the taking of eggs at Ostional and establish a management system to control both the collecting and marketing of eggs. At that time the human population of Ostional had grown to about a hundred. The legal right to harvest ridley eggs was vested in a local economic development association, the Asociación de Desarrollo Integral de Ostional (ADIO).

The University of Costa Rica was given the task of carrying out scientific studies to monitor the *arribadas* and the progress of the harvest system. Doug Robinson and his students had the legal responsibility for preparing an annual plan and reviewing progress of the association and the harvest. They also were to provide the scientific underpinning for the project. The ADIO was to administer the project, control the harvest, guard the turtles, protect the beach, and control the sale and marketing of the eggs. At first, the system seemed to function well.

All of the people who were residents of Ostional in 1987 could become members of ADIO. A child could join when he or she turned fifteen (today membership is limited to anyone who is fifteen or older and has been a resident of Ostional for at least 3 months). In the morning of the second day of an *arribada,* the egg collectors went to work. They were organized into work groups, each group having its own section of beach to cover. The men found the nests with their heels and marked them with a tag or flag. The women and girls followed along and dug up the eggs, put them into large sacks, and took them down to the ocean to wash them. Then they carried the eggs to a central location, where they placed them, two hundred at a time, into small plastic bags stamped with the name and seal of ADIO. This process only took place for the first 36 hours of the *arribada.* Then the sacks were loaded on trucks and sent off to market. At the same time other members of ADIO worked as guards to keep outsiders from poaching eggs. They would continue this vigilance throughout the *arribada* and during the time between *arribadas* as well.

Profits from the egg harvest were distributed as salaries to ADIO members (70%) and to community development projects and ADIO expenses, including paying the biologist's salary (30%). This system hinted at problems for the future, since the biologist served at the pleasure of ADIO, making it difficult for him to provide an independent estimate of the impact of the harvest on the turtles and their reproductive success. Nevertheless, initially and for some years, the system was broadly supported by the local community.

Lisa Campbell, now a faculty member at Duke University, carried out the most comprehensive study of the socioeconomic aspects of the harvest system in 1994–1995. She found that 70 percent of households identified egg collecting as their main activity and that their salaries exceeded what they could earn in other local activities. Benefits from ADIO were well distributed in the village, although some families with many members earned more than the average and the people who had the permits to distribute and sell eggs earned much more. Other households that did not have ADIO members were also supportive of the system because they earned money from their stores or from doing business with the members, plus the level of violence was lower than before the association got control of the egg harvest. People agreed that there was a need to diversify the local economy to reduce dependence on the egg harvest because that system relied on the arrival of turtles only a few months out of the year. During other times there was little generation of income in the village.

The only other potential source of income for the villagers was ecotourism. The rest of the country was seeing a boom in such activity, and the people at Ostional saw it as a new source of money. However, those three rivers and the deep Rio Nosara to the south made access impossible but for the most intrepid ecotourists. Foreign developers were active in Nosara, but few people wanted to invest in Ostional, since members of ADIO were still hostile to outsiders.

There were only a couple of "hotels" in Ostional when Lisa conducted her study, and their average occupancy rate was only about 12 percent. Still, the families that ran the two *cabinas* in the village in 1995 earned four to seven times what they could have earned in the egg harvest. That led to resentment among many people in the village. Many egg collectors were suspicious of anyone who would let foreign tourists come to the beach, because there were few English-speaking residents who could be guides. People were afraid that tourists might disrupt the turtles, although olive ridleys in an *arribada* seem pretty insensitive to the presence of people, and that more tourists could bring in more foreign presence in the village, not to mention an increased scrutiny of the harvest.

The Middle Years

Unfortunately, Doug Robinson passed away at too young an age and the scientific leadership of the project at Ostional was lost. Some people say that one person cannot make a difference, but in this case Doug was critical to the Ostional experiment, and without his vision and leadership it soon went astray. The idea had been to develop a program of sustainable development and scientific management of the turtle population that nested at Ostional. There would be sustainable use of wildlife resources (turtle eggs) and community-based conservation, that is, the villagers would be sure that the turtle population was protected and sustained. Egg collecting would lead to additional economic development, and eventually those other economic drivers would reduce the demand for eggs not only in Ostional but also in the country as a whole. Sadly it was not to be. The project at Ostional became captive to economic and cultural pressures and was no longer guided by sound science. The scientific studies to test the hypothesis that this harvest system was helpful to the hatchling productivity were never done.

Lisa Campbell discovered that members of ADIO were strong supporters of the system and that there were extensive economic benefits. At first ADIO contracted the distribution of eggs to an outside vendor but later took it over directly. Profits increased and funded construction of a health center, a house for schoolteachers, the ADIO office, an egg-packing building, a laboratory for the University of Costa Rica researchers, and school remodeling.

However, the driving force for management of the egg harvest system was the economic benefit of ADIO and not the biological benefit of the turtles. As the number of members of ADIO increased, the rate of harvest per person remained the same. Therefore, the take of eggs increased from tens of thousands in the early days (providing a few thousands of dollars to fewer than 100 members) to 5,278,800 eggs in 2008 (providing $347,289 to over 600 members).

The real biological questions were: What was the impact of the harvest system on the biology of the turtles? Was the harvest increasing overall hatching success by removing clutches that would be destroyed by turtles nesting on subsequent days? Were there more turtles than in previous years? In fact, how many turtles were there? No one was answering those questions.

It is notoriously hard to count turtles during an *arribada*. There are so many turtles milling around, coming and going, and starting to dig a nest then leaving to come back later and dig again that no one had an accurate way to count them. The protocol was to count turtles in a defined section of the beach. If the turtles shifted to another section during a given *arribada*, the biologists still counted turtles on the defined section. So the estimates of how many turtles participated in the *arribadas* were totally unreliable. We don't know how many olive ridleys nested at Ostional during those years.

The refuge biologist was controlled by ADIO and did not have the independence to regulate the harvest system. The Ministry of the Environment and Energy (MINAE) did not have a ranger or other representative in the wildlife refuge. It also had no permanent headquarters there. So rangers only visited the area on occasion, perhaps for only some of the *arribadas*. The members of ADIO policed themselves. In this state of confusion conflict developed between members of the community and the biologists from ADIO and the University of Costa Rica.

During this time in the mid- to late 1990s, Anny Chaves, one of Robinson's students, took over the Sea Turtle Laboratory at the University of Costa Rica (UCR). Anny inherited the project under difficult circumstances. What had been simmering feuds between biologists and community members, and among community members, erupted into a series of unfortunate events, including a nighttime attack on the official ADIO/UCR biologist and the burning of the biological station. Neither of those crimes was ever solved.

Recent Years

After 2000 the situation improved, as MINAE took a stronger hand. A new director of the wildlife refuge made considerable progress in bringing order to chaos. Carlos Mario Orrego is a very bright, competent, and concerned veterinarian trained in Colombia and Costa Rica. He was a veterinary student and head of a student organization in Colombia when some paramilitaries came to his house looking for him. He was not home, but they gunned down his brothers instead. Carlos and his family left their country and became political refugees in Costa Rica and the United States. Carlos obtained a master of science degree in wildlife management and conservation from the National University in Costa Rica, became a citizen, and was appointed director of the Ostional Wildlife Refuge.

Carlos undertook a long process of winning acceptance from the local community and establishing a full-time presence for MINAE at Ostional. In 2003 MINAE assumed leadership of the process of participatory environmental management at Ostional and called a meeting of all the stakeholders in the Ostional Wildlife Refuge. These included ADIO, UCR, INCOPESCA (the government fisheries management agency), the local municipalities, fishermen, and other local residents. A consensus was developed and the stakeholders created a co-management commission called CIMACO (Ostional National Wildlife Refuge Inter-institutional Consulting Council) that became legally established by executive decree in 2008. Carlos also raised funds to pay for constructing a headquarters and ranger station for the wildlife refuge. Each month the commission would meet at the refuge facility and discuss issues such as protection and control of the harvest, environmental education, ecotourism, legislation, and management. Finally, progress was being made and order established in the refuge. This required patience and hard work because the leadership of ADIO did not want to relinquish its vigilante hold on the community and the operation of the refuge and egg harvest. They had gotten used to full community-based conservation. They were in charge and wanted to set the rules, legal niceties and regulations notwithstanding. It turned out that Carlos was quite the diplomat with just the right blend of patience and perseverance to succeed.

In 2003 a local group of twenty people formed an ecotourism cooperative, and Carlos helped them get certified as official guides. Many young people were enthusiastic about this new form of employment, and they had much better English language skills than their parents. The refuge hosted meetings on recycling, solar energy, handcrafts, orchid culture, and alternative fishing techniques. Carlos encouraged scientific research and worked with Roldán Valverde, a biologist at Southeast Louisiana State who grew up in Costa Rica, to establish an accurate method for counting nesting olive ridleys. This study is being conducted in Costa Rica, Mexico, Nicaragua, Panama, and India. It is basic to any attempt to ascertain the impact of egg harvest on olive ridley populations. A lot of progress has been made recently, but much remains to be done.

The Stain on Conservation

Ostional has been held out as a jewel of sustainable management of sea turtle populations. Costa Ricans and people around the world refer to it as a beacon of hope for maintaining a balance between local people and their sea turtles. However, as Lisa Campbell points out, while the economic development aspects of the project have succeeded and most local residents support the project, that does not translate into success in terms of conservation. We do not know if the egg harvest has improved hatching success at Ostional, because that hypothesis has not been tested with a rigorous scientific study. We do not have good historic data on the numbers of turtles in *arribadas* from 1990 to 2005, because the counts were not done in a rigorous scientific manner and were not done in the same place and in the same way for every *arribada*. Some *arribadas* were not even counted; they were estimated by members of ADIO. That anecdotal data suggested that numbers of turtles were stable or increasing, but the data were unreliable, so we do not even know the size of the Ostional population.

In addition, it is to ADIO's economic benefit that the number of turtles increases, because that justifies the increase in the numbers of eggs that they collect. How can you trust the fox to count the eggs in the chicken coop? You need an independent agency to do that. The UCR biologist is paid by ADIO. Where is the independence there? An outside biologist supervised by MINAE needs to do these assessments. Some of that is being done now by Roldán Valverde and Carlos Orrego, but the data will not be in for some time.

On top of all of these problems is the fact that the egg harvest at Ostional has not lowered the demand for sea turtle eggs in Costa Rica. If anything, it has increased that demand. If you go to any beach in Guanacaste that is not in a park, you will find that poachers are on the beach and that they take all or almost all of the eggs. That is true at Nombre de Jesús north of Las Baulas Park, Playa Junquillal, Playa Cabuyal, and Playa Panama. I already described the situation on Tortuguero and Pacuare beaches on the Caribbean coast. In supermarkets and bars in San José you will find packets with the ADIO label that contain green turtle eggs and leatherback eggs. Now the former might be mistaken for ridley eggs by the uninformed, but no one can mistake a leatherback egg for an olive ridley egg. Are these bags counterfeit? Are they real but stolen or diverted from Ostional? Where do they come from? Where do the eggs come from? No matter what the answer, the consequences are clear. The harvest of eggs at Ostional increases demand for sea turtle eggs and creates a cover for people to poach eggs from other beaches around Costa Rica. It has not solved poaching but has made it worse.

If Doug Robinson had lived longer perhaps we would have understood the population effects of the egg harvest at Ostional. Perhaps Doug would have seen that the eggs from Ostional were being used as cover for eggs poached elsewhere and were fueling a market for sea turtle eggs. Instead, we have the Ostional harvest continuing, growing, and doing an unknown amount of damage. I cringe when people talk of Ostional as a model for the rest of the world. This is what we have learned from the model: we should not encourage taking of eggs or adult turtles anywhere in the world unless we document that it fosters sea turtle conservation and does not produce markets for turtle eggs from other locations. The egg harvest at Ostional fails on the second count, and perhaps the first, and should be ended.

12

GLOBAL WARMING: RISING SEAS, LOST BEACHES, AND GENDERS

In the past, the earth seems to have warmed or cooled over hundreds of years. Sea turtles presumably responded by shifting nesting beaches as the oceans swelled. Many of the changes were probably very slow, and there were plenty of open beaches back then. Today we have a few protected beaches and many developed beaches. If oceans rise and flood protected beaches, where will the turtles go? The beaches of Florida, the Carolinas, and Georgia, where green turtles, loggerheads, and leatherbacks nest, are the last pristine—or at least suitable—beaches available. I've heard people say they can move north. Perhaps they already are. Thirty years ago it was pretty uncommon for loggerhead turtles to nest in North Carolina, and it was a big surprise when a few nested in Virginia. Now nesting is more common in both states, although still very rare in Virginia. In 2009, green turtles and a few leatherbacks nested in North Carolina and a green turtle even nested in Virginia. Delaware has done a good job protecting beaches, but first sea turtles have to survive years of nesting along Virginia and Maryland's overdeveloped coasts. And God help them if they get to my home state of New Jersey.

Changes in My Lifetime

In 1971, I was a postdoc with David Gates, and during a lab meeting David was excited to show us the latest data that C. D. Keeling had collected on CO_2 levels on Mauna Loa volcano in Hawaii. Keeling had been measuring CO_2 there since 1958, and the plot of CO_2 versus time, now called the Keeling curve, recorded the change in atmospheric CO_2 since that time. The value was about 323 parts per million (ppm). The interesting thing was that it was about 317 ppm in 1958, so it was going up. All of us postdocs started to get excited and said we thought this meant there was a problem, what with all of that heat balance of the earth business we had learned about. David, being the wise and distinguished professor, said: "Now don't get carried away. This looks like a trend, but we can't be sure. It might be normal variation. But if it continues going up, then we will have a real problem."

Eighteen years later, in 1989, I was a professor at Drexel University. Frank Paladino, Mike O'Connor, and I were on Tortuguero beach in Costa Rica to measure metabolic rates of leatherback turtles. We collected the expired air from ten breaths into a big rubber weather balloon, measured the time for those breaths, and took the bag of gas back to the primitive laboratory at the Casa Verde field station. There we used a CO_2 meter and an O_2 meter to measure the gas content of the breaths and compute the metabolic rate of the turtle. To see if our equipment was accurate, we measured the atmospheric CO_2. It should be 323 ppm, plus or minus 10 ppm, I figured, if Keeling's data had kept rising or fell back some (it had gone up 7 ppm in 13 years, so 10 ppm in 18 years seemed reasonable). But the number I kept getting was 350 ppm. I was sure it was too high, but we forged on with our instruments, using that as a guide. We could recalibrate to the real, lower, number later. When I got back to

Drexel, I looked up the Keeling curves. The value for CO_2 from Mauna Loa was 350 to 353 ppm! The CO_2 meter at the beach was correct. It was sobering. Trouble was brewing.

In 2010 the CO_2 reading from Mauna Loa was 390 ppm. The levels are rising faster than ever.

Effects on Sea Turtles

One of the obvious effects of global warming on sea turtles is that their nesting beaches are getting hotter. Hotter beaches mean higher nest temperatures, which produce more females. Playa Grande in Costa Rica has been producing primarily female leatherback hatchlings for many years. We know that from the work of Annette Sieg and Christopher Binckley, who both earned graduate degrees at Drexel University: Binckley an MS in 1996 and Sieg a PhD in 2010. Binckley, now a faculty member at Arcadia University, did the experiments to determine the effect of temperature on temperature-dependent sex determination in Pacific leatherbacks.

Sieg and Binckley used data on beach temperatures from 1993 to 2007 and weather data from nearby Santa Cruz to formulate a mathematical model to predict nest temperature from weather data. Then they took weather data from 1950 to 2007 to calculate nest temperatures and to predict sex ratios. In most years the sex ratio was highly biased toward females—more than 73 percent. In the few years when there was a strong La Niña and it rained a lot, the sex ratio tipped toward males.

Global warming is also threatening sea turtles' food supply. Colin Limpus, one of the foremost field biologists studying sea turtles, was the first to find a link between the El Niño–La Niña phenomenon in the Pacific Ocean and sea turtle nesting. During El Niño the surface waters of the Pacific warm up and the upwelling of cold waters stops. The rising cold bottom waters bring nutrients to the surface and fuel algal growth, which drives the food chain. Warm El Niño water shuts that off. La Niña means cooler waters than normal and increases in productivity. Numbers of green turtles that will nest in Australia can be predicted by ocean temperature a year ahead of time. The warm temperatures of El Niño suppress nesting activity by reducing food for the turtles. Leatherback turtles in the Atlantic Ocean return to lay eggs every 2 years at St. Croix in the U.S. Virgin Islands, while they return to lay eggs every 3.7 years at Las Baulas Park on the Pacific coast of Costa Rica. Leatherbacks are larger at St. Croix and lay an average of eighty-five eggs per clutch. At Las Baulas the smaller leatherbacks lay clutches of sixty-five eggs. El Niño is the cause.

Bryan Wallace determined the reasons for this difference as part of his PhD at Drexel in 2007. Bryan is a bright and high-energy guy. He used a new technique to measure the metabolic rate of free-swimming leatherbacks during the time they were at sea between nesting bouts (they lay seven clutches in a season). He injected water containing stable (nonradioactive) isotopes of hydrogen and oxygen into the turtles when they were on the beach and measured the amount of the isotopes left in the blood when they returned to nest. The rate of washout of the isotopes was a measure of the metabolic rate. At the urging of Frank Paladino, Bryan made the technique work by using labeled water (H_2O) that had very high amounts of oxygen 18 and hydrogen 2 in it.

Then Bryan computed the energy cost of foraging and producing eggs for a leatherback. A leatherback could return to nest again in as little as a year if it could find enough to eat. If food was scarce, then it had to wait longer to return because the energy used looking for food took away from the energy that could be stored as fat to make eggs.

Data from George Shillinger, a creative graduate student at

Stanford who put satellite transmitters on leatherbacks at Las Baulas, indicated that leatherbacks in the Pacific did not have a particular place to feed. They tend to wander in the equatorial and South Pacific and along convergence zones, where food is abundant. By using stable isotopes of nitrogen (N_2) and carbon (C) in turtle tissue, Wallace confirmed that the main source of food for Pacific leatherbacks was the upwellings of cold, nutrient-rich waters that fed those convergence zones.

At the same time Vince Saba, who had demonstrated great quantitative and people skills at Las Baulas while an MS student at Drexel, did some sophisticated mathematical analyses of the relationship of El Niño to leatherback nesting ecology for his PhD at the Virginia Institute of Marine Science. He demonstrated that renesting probabilities of leatherbacks at Las Baulas were tied to the El Niño–La Niña phenomenon. Cool La Niña events corresponded with a higher remigration probability, and warm El Niño events corresponded with a lower remigration probability. Vince followed up that work with a more sophisticated analysis of the relationship of leatherback nesting with ocean productivity around the world. He found that the difference in reproductive output of leatherbacks in different areas was caused by natural differences in climate variability compounded by the influence of global warming on the Pacific Ocean cycles of El Niño and La Niña. So it is pretty clear that climate affects many aspects of sea turtle life and that global warming is having a profound effect on all of them.

Turtle Nesting Beaches

The famous green turtle nesting beach at Tortuguero is 20 miles long. It is a barrier beach with a lagoon and river behind it. The sands are a meter above sea level. As sea level rises, the current beach may become sea bottom and the wetland swamp behind the lagoon may become a beach. Fortunately, the wetland swamp is a wildlife refuge and park, and if the turtles are lucky the new beach will be suitable for nesting. What are the chances it will be? No one knows.

On the other side of Costa Rica, at Playa Grande lies the peninsula that separates the open beach from a large mangrove estuary. It is also only about a meter above sea level. As glaciers slide off into the ocean and sections of West Antarctica melt, the Pacific is expected to rise, so that the beach at Playa Grande will also be sea bottom. What will the people who have built houses and hotels on the land behind the beach do to try to "save" the beach? Will the water rush in over a couple of weeks or a year, or more gradually? Will the beach continue to reestablish itself as it overruns the estuary? Will all of the leatherback eggs die because they are washed by rising seawater? There are a lot of unanswered questions for both people and turtles.

Florida beaches are, in general, occupied by houses, condominiums, and hotels. Many have sea walls behind the beach to protect their property from crashing waves. There is no place for new beaches to become established when the sea rises. Loggerheads that now nest on those beaches will find only water meeting wall (the way many parts of Cancún are today at high tide). The Archie Carr National Wildlife Refuge, which stretches about 21 miles from Melbourne Beach to Wabasso Beach along Florida's east coast, protects 25 to 35 percent of all loggerhead and green turtle nests in the United States and some leatherbacks as well. Up to 15,000 turtles nest at Archie Carr in a good year, and most of the refuge provides natural habitat with few or no artificial lights. This is the best hope for Florida sea turtles because as sea level rises, new beachfront could develop where dunes exist today. Hurricanes Frances and Jean in September 2004 moved this beach back 70 feet in places and destroyed some houses. But the beach recovered, and turtles still nest there.

An Alternative Future

You have probably seen one of those science fiction movies or TV shows in which someone goes back in time and changes history and when the protagonists return they find themselves in a different world. Imagine if we could go back to 1950 and convince people to change their ways and avoid the dangers with which global warming now threatens sea turtles and us. It is too bad that can't happen. But there is something we can do to create an alternative future even now. Much of what was discussed in this chapter assumes that CO_2 would continue to increase and might reach 600–700 ppm by 2050 and that the earth would continue to warm into the next century. It is true that considerable warming is already built-up in the climate system and the oceans will rise. However, we can change the future. We have to reduce CO_2 levels to at most 350 ppm by 2050 to save our planet and its sea turtles. How can we do that? Here are a few suggestions:

1. Work to eliminate the burning of coal from power plants that do not recapture the CO_2 produced. That is the surest way to reduce CO_2 levels to 350 ppm.
2. Drive a hybrid car that gets 35 to 50 miles per gallon (mpg).
3. Use public transport. If you live in a city, buses and trains are going where you need to go anyway, so get on board and eliminate all of that CO_2 your car produces.
4. Use fluorescent light bulbs. Fluorescent light bulbs use 25 percent of the energy of regular incandescent light bulbs.
5. Turn off unused lights. We all tend to leave lights on when we go from room to room. Even if they are fluorescent they still burn energy and produce CO_2.

6. Upgrade your home's insulation. Replace old windows with energy-efficient new ones. Check your house walls and attic, and add insulation if it is needed.
7. Install solar panels on your roof. Soon you will be *making* money by selling excess power to the power grid *and* saving a bundle on your electric bill.
8. Build a windmill. Windmills can provide a lot of electricity if they are placed in the right locations.

Two other big things that need to be done are:

1. Build nuclear power plants. France gets 80 percent of its power from nuclear power plants, so why can't other nations? In the United States, and Pennsylvania in particular, there is a lot of fear of nuclear power. No one wants a meltdown. But nuclear power plants can be built correctly and safely. That is really the only practical way that we can reduce our CO_2 emissions to safe levels by 2050, unless we stop using electricity.
2. Engage with China and India. China now produces more CO_2 than the United States and has a booming economy. There are 300 million people living in China's cities, and there are 125 cities of a million or more people. China has several cities that most people in the United States and Europe have never even heard of that are bigger than Chicago. There are 900 million people down on the farm using little energy, but they want to live in cities, where there are more jobs. So you can easily see that there will be a lot more CO_2 produced in China if something does not change.

India is also growing and producing huge amounts of CO_2. We cannot solve the global warming problem unless these nations get aboard the climate train. There is interest in China to solve this problem. The United States needs to take the lead and to be willing to partner with China and India to find a solution. You can help by talking to your colleagues and friends from those countries or others who are of Chinese or Indian descent. They have relatives there, and word will get back to them. Do what you can, for humanity and for the turtles.

13

THE GREAT TURTLE RACE: A NEW APPROACH TO CONSERVATION EDUCATION

The Great Turtle Race (www.greatturtlerace.org) in 2007 was one of the most interesting and effective conservation awareness campaigns ever conducted on the Internet. Media coverage of the event reached more than 100 million people around the world. And it all began with a graduate student and an Internet entrepreneur talking about how to translate scientific information into a web-based strategy to save sea turtles. The graduate student, George Shillinger (now a PhD), was studying at Stanford University. His dissertation research involved satellite tracking of leatherbacks. The entrepreneur, Mark Breier, was and is a marketing genius. He is the author of the best-selling book *The 10-Second Internet Manager: Survive, Thrive, and Drive Your Company in the Information Age.*

George and Mark had been introduced by Rod Mast of Conservation International (CI). In 2005 George described his interest in using his tracks of leatherbacks swimming from Costa Rica to the Galápagos Islands to raise conservation awareness and fund turtle conservation. George was borrowing ideas from some bird researchers in the South Pacific and United Kingdom who had put flight paths of albatrosses on the Internet as part of an education effort. Another, unaffiliated, website hosted betting—yes, people actually bet on which bird would get across the ocean first!

Mark and George examined the turtle migration routes from Costa Rica to the Galápagos feeding grounds, agreed that it "looked like a race," and started brainstorming. Putting the tracks on the Internet seemed fine, but what about treating it like a yacht race for millionaires (with a catchphrase along the lines of, "trading flippers for skippers"), a betting race (a la Las Vegas), or a corporate sponsor race (like NASCAR)? No doubt was left in their minds: the migration routes of leatherbacks could be an educational and marketing tool that taught people about biology and the plight of sea turtles.

A Unique Partnership

Mark was no small-time entrepreneur. In addition to his author credentials, he had been vice president of marketing at Amazon.com at the moment it really took off. He was also president and chief executive officer of Beyond.com and was credited with transforming that company into a top-ten e-commerce site. When I met him he was vice president of Plantronics, the company that makes Bluetooth headsets.

Mark got his start in brand marketing with Kraft Foods and General Mills. But he's not a stereotypical "marketing suit." He's a down-to-earth, hands-on type of person. For example, when he became vice president of marketing for Cinnabon Corporation, he worked in a Cinnabon store for a week to learn how to make those delicious treats and to see how the business worked on the front lines.

George, while working on his PhD at Stanford University's Hopkins Marine Laboratory, was also a part-time development officer for CI. Before going back to school, George worked for CI for many years. As a member of the Andean Regional Program of CI, he helped write the proposal that led to the Eastern Tropical Pacific Seascape project, which now brings together several countries to promote sustainable development of marine resources between the South American mainland and the Galápagos Islands. In addition to a bachelor's degree in biology from the University of Pennsylvania, George had master's degrees in business administration from Yale University and ecology and evolutionary biology from Stanford. So George knew a thing or two about mixing the worlds of biology and business.

To be successful, George and Mark's idea had to be both simple and catchy. Find a corporate sponsor for each turtle, let all the rest of the world pick a favorite, and watch the race for a few weeks or so. The first turtle to cross the finish line (the outskirts of the Galápagos feeding grounds) wins.

Together Mark and George began to explore sponsorship opportunities for "The Race." They wanted corporations—which are seldom given as much credit as they deserve for their contributions to good causes—to provide money that would be used for leatherback conservation. After an unsuccessful exploratory meeting with eBay, they secured Yahoo as a lead sponsor and host for the race. That was good news, but they needed a nonprofit organization to receive the funds from the corporate sponsors and put the funds to good use. George, Mark, and Rod Mast promoted the race to CI's Development Team and Marine Conservation Program. Mark is a member of CI's Chairman's Council, and Rod leads the sea turtle group at CI and is vice president of CI-Sojourns (educational trips for donors). Rod's work includes taking people like Harrison Ford and Rob Walton (of Wal-Mart fame), around the world to see all the places, ecosystems, plants, and animals that CI is trying to save. These influential people then assist CI in their efforts. Rod helped George take it up the line to the CI Development Office, but like many good ideas, it had its day in court and was turned down.

So there they were: Mark with his business skills and George with his conservation expertise. Between them they had the biology and the Internet marketing figured out. They had Rod Mast on board but nowhere to go.

Let's Do Lunch

During November 2005 I gave a talk on sea turtles that George arranged at the Bullis Charter School in Los Altos, California. George had been working with Mark Breier and two other board members (Kris and Ken Moore) and the school's faculty to involve the school in international leatherback conservation and education projects in Costa Rica. After a very interesting session with bright and engaging students from the first to sixth grades, I was talking to Wanny Hersey, the principal, and I invited her to bring a group of students down to see the turtles. Of course, I had given that invitation to many groups and it was usually one of those "let's do lunch" offers. Nothing ever comes of it.

This time it was different. Two weeks later Wanny called me and said they would like to take me up on the offer and have some of the older students come to Costa Rica soon, in February. So with a lot of hustle on both ends, we set up the trip. I was a little unsure about spending a week on a bus with twenty-four sixth graders, but the children were great, and teachers Lisa Stone and Jessica Lura kept them occupied and well behaved. Lisa and Jessica showed a wonderful mix of enthusiasm and knowledge of both the turtles and Costa Rica. The visit from this school has since become an annual event, but that first trip also indirectly led to fruition for the race.

In April I was back at Bullis to meet with the children to hear their report to the school's board of directors on their trip. It had been the first time that I had hosted an elementary school group to see leatherbacks at Las Baulas Park, and I was eager to see what they had learned. I was impressed, to say the least. That night, after the children gave their report, I was sitting and having a glass of wine with Mark in his kitchen and I asked him about the race. Had anything come of it?

After Mark told me the story of their failure to woo the development staff at CI, I simply said, "Let's do it." He said, "What?" Again I said, "Let's do it." As president of the nonprofit organization called the Leatherback Trust, I told him we would sponsor the race if he was willing to continue to help make it happen. So we shook on it and called George. He quickly agreed. I started pulling my people together, telling them that the race would run in the spring of 2007. George and Mark were convinced that the leatherback migration had all the elements of a good race: varied speeds, as some turtles stayed around the nesting beach while others sped off toward their foraging grounds; fun facts such as record-setting dives by some of the turtles; and a variety of turtles—from first-time nesters to seasoned mothers. The racetrack was vast—the eastern Pacific Ocean. It held many hazards, from thousands of fishing hooks, long nets, floating plastic that looked like the leatherbacks' favorite food (jellyfish), and strong currents.

Scientists often use satellite telemetry to study movements and behavior of sea turtles and other animals. Current off-the-shelf satellite transmitters report position, water temperature, swimming speed, diving depth, time, and other data. We wanted to convey to children, their parents, and the general public the excitement of all that has been discovered about leatherbacks through telemetry. We also wanted to inform them about the plight of leatherback turtles in the Pacific Ocean, and that the Pacific population was in the midst of a devastating decline.

Ten Turtles

By June of 2006 we were well on our way. Mark, George, Rod, and I formed a committee to organize the race. My handshake deal was sealed when the Leatherback Trust board enthusiastically endorsed the idea of "The Great Turtle Race" and authorized me to move forward full speed ahead. We acquired the Internet domain names www.greatturtlerace.com and www.greatturtlerace.org, but truthfully I had no idea of where to go from there. I had one of those "crisis moments" when a person says, "What have I gotten myself into?" (I think my three partners felt the same!) I knew that our small nonprofit group was truly a fledgling effort at the time. I knew that the Trust could not front more than a few thousand dollars for the project. It seemed to me that we needed more corporate sponsors—and soon—to get the process going. Mark set a price: for $25,000 a corporate sponsor would pay for one turtle. Of course, paying for one turtle meant paying for one transmitter and for a share of the expedition required to find the turtles and attach the transmitters, and also for data acquisition and analysis. There would, Mark figured, be funds left over to set up and manage the website and to fund other efforts.

Not having graduated from the Yale or Stanford business schools, I relied on Mark and George to do most of the real thinking. Our conversations were by e-mail, phone, or later, when the group expanded, by conference call. It would have been impossible to pull this project off even 5 or 10 years earlier because the capability for long-range, real-time discussion via electrons just was not there and our team was scattered across the country.

Mark suggested that we get a working team together, agree on the general idea, and then nail down several of the key elements, the latter including a better understanding of the finances by answering some key questions: What are the actual costs of the different items—transmitters, tracking, data management, website, etc.? What revenues could we expect? Should we offer just ten turtles at $25,000 each? What is the prospect for philanthropic participation by companies and media? How about naming the turtles after the sponsors, but in a humorous way?

Mark wondered if the race could be "tape-delayed," as TV programs are, particularly for West Coast broadcasts. Maybe the event could be "weeks-delayed" to package and polish the race before it went live. As it turned out, that was the model that we would use. George, however, had some concerns about the credibility issue if the race was not in real time. That was something that we would have to resolve as the planning process progressed. Bottom line: we were starting to develop a workable plan.

By July we were making real progress. Rod was very excited and said that he thought CI would still want to be a partner in the race. Mark then talked to Murray Gaylord, vice president for brand marketing at Yahoo. We received the good news that Yahoo was still interested in the concept. Murray had run the Ad Council (nonprofit ads on TV) for many years and thought this idea had wonderful potential. We sent Murray materials on leatherback turtles, examples of tracks, backgrounds on the nonprofits involved, and a ballpark prospectus for Yahoo involvement. We then began planning a presentation for Yahoo.

About this time George said, "Things are going pretty fast, but before we go too far down the road we need to have the TOPP [Tagging of Pacific Pelagics] project at Stanford involved in the race." The TOPP project, led by Barbara Block, was putting transmitters on all of the large animals in the northern Pacific Ocean to measure their movements, behavior, and home ranges. The project also used the data from the transmittered animals to measure the physical parameters of the oceans.

Our specific plan was to put satellite transmitters on leatherbacks in Costa Rica and follow them to the Galápagos Islands—it seemed a natural fit for TOPP. We needed transmitters, we needed to deploy them, and we needed the data analysis team. George said he could handle the transmitters and deployment and that perhaps the TOPP programming team could do the rest.

August was soon upon us and things were heating up. Mark put together the start of a PowerPoint presentation for Yahoo while Rod Mast sent one that his team put together. We merged the two into a good briefing presentation. The cart was a little ahead of the horse at this point, as we did not yet have formal agreements with CI and TOPP to participate in the project. And George was volunteering extensive time to it while awaiting the formal approval of Barbara Block, his major professor (a major professor oversees her graduate students).

I suppose the lesson from this story is that you can't always wait for all of the paperwork to be done before you go ahead with a good idea. We reasoned that the final approvals would be more likely if Yahoo was on board, and Yahoo did not need to be troubled with details about which organizations had formally signed on to the race. As far as Yahoo was concerned, the three groups were all on board, and as far as CI and TOPP were concerned Yahoo would soon be on board. In other words, we bent a few rules for a good cause.

By early August we worked out details of TOPP's participation and briefed Yahoo. The meeting came off without a hitch, and Yahoo was on board as overall sponsor. The next day Mark announced that Plantronics had agreed to be the first turtle sponsor, and he brought in McCann Erickson, a marketing firm from Salt Lake City. Then Mark secured Dreyer's Ice Cream (known as Edy's on the East Coast of the United States) as a sponsor, and I committed Drexel University, figuring I would check with my boss, President Constantine "Taki" Papadakis, later. He was never one to miss an entrepreneurial opportunity, so I thought the university would find the money somewhere—out of my own endowment at Drexel, if necessary. I did eventually get Taki's enthusiastic support in the fall, but he was pretty shrewd—and made my endowment pay for one-half of the turtle. Our team was growing, too, as Lisa Bailey, Brian Hutchinson, and Vinnie Wishrad from CI got involved. Lisa did an awesome job on media relations throughout the project, Brian took care of turtle stuff at CI, and Vinnie oversaw CI's web team.

Moving into High Gear

By September we had only 4 months to meet our January deadline for testing the website. All the experts said we were cutting it too close. We did not even have a concept design for the site yet. We were also just beginning to think about a necessity—a celebrity host. The folks at CI took on that task. We also needed to order transmitters, get animal care committee approval from our universities, and obtain permits from the Ministry of the Environment and Energy in Costa Rica. The minister was excited to be involved because he loved technology, and Rotney Piedra, the director of Las Baulas Park, was especially excited because he and his rangers would be directly involved in deploying transmitters. At the end of September we began a series of conference calls to get all participants on the phone at one time. By now we were up to about ten people and needed to get everyone on the same page and keep them focused on the Great Turtle Race as their top priority.

We received a horrible surprise when Murray suddenly stepped out of the picture. The new people we talked to at

Yahoo drafted an agreement that would have *us* pay Yahoo for the website use. In addition, the donation we were counting on from Yahoo was gone. Needless to say, that put a huge damper on the race. However, Mark continued to work the situation, and the final agreement between the Trust and Yahoo called for Yahoo to host the website, to get a free turtle, and for the Leatherback Trust to pay $75,000 for ads on Yahoo to cover the cost of hosting the site. I was shocked! Mark said it was the best deal we could get. I really wanted to balk, but Mark had better business sense and said we should take it. Without Yahoo, he reasoned, we could not get our other sponsors. We all swallowed hard and agreed.

So we got the prestige of having the Yahoo connection, and gave Yahoo a turtle for free. It was never clear to me how many ads Yahoo ran, but in the end they never billed the Leatherback Trust for any of them. A couple of years later I verified with Yahoo that the Trust could consider the $75,000 worth of ads as a donation. Perhaps someone had intervened? Perhaps Yahoo's heart grew two sizes? We'll probably never know, but I'm grateful nonetheless.

October was a busy month. Mark developed a storyboard version of a website for a meeting with Yahoo. Looking at it now, it is amazing how much of what he sketched out actually foretold what the finished website would look like. He developed the home page concept that had a picture of the race and tabs to Adopt a Turtle, Meet the Turtles, Learn More, and Merchandise pages. The Meet the Turtles page was pretty close to what finally made it onto the site, while the Learn More page evolved into the Sea Turtle School page.

Mark also made the first clear statement of our goals: to raise awareness regarding saving sea turtles, raise money to cover costs, make sponsors successful, contribute to science, and prove concepts for future races. Here we were in the fifth month of the project and we finally had a full set of specific goals! I was particularly impressed with Mark's insistence that we make the sponsors successful. Having good business sense, Mark knew that sponsors might be giving an initial donation to save turtles, but they would be likely to participate in the future conservation campaigns only if they gained something from the experience.

When all the dust settled, Mark had brought in Plantronics, Dreyer's, and Travelocity. George brought in the Offield Center for Billfish Studies, Bullis Charter School (with a little help from me), West Marine, and GITI Tires. I brought in Drexel. We had two turtles to go. The CI people were working to secure the right celebrity spokesperson but were not yet successful in getting anyone to agree.

In January the team was hard at work creating the website but still did not have any transmitters on turtles. George headed to Costa Rica in late January and, with the help of Bryan Wallace and Rotney Piedra and his rangers, started putting the satellite transmitters on the leatherbacks. He finished the job of setting out eleven transmitters (one extra for good luck!) on February 15, the same day that the website design was completed.

The Celebrity

The last month or so was crazy, with so many last-minute items to complete and so little time. Luckily we were joined by a volunteer, Brooke Glidden, who played a key role in our final success. Calling northern New Jersey home, she is the mother of Johnny, a major leatherback turtle advocate. Johnny was nine that year, and he had been giving talks in local schools about leatherbacks for a couple of years. George had gotten the two of them down to Las Baulas, and they fell in love with the turtles and the park. Johnny and his mom made a full-size leatherback model, "Leopold," out of fabric and stuffing, and Johnny would take it to different schools and tell the children about saving leatherbacks. He even made a video and put it up on Yahoo.com!

We had nine sponsors in hand for eleven leatherbacks. Brooke suggested that we add the Life Science Secondary School in New York to complete our elementary-to-university trio (in age order: Bullis Charter School, Life Science School, and Drexel University) and to generate interest in the New York media market. That left one turtle and still no celebrity sponsor. It turned out that Brooke would fill that hole.

Brooke knew a producer on the award-winning and extremely popular *Colbert Report* on Comedy Central and talked to him about naming a turtle in Stephen Colbert's honor in exchange for coverage on the show. She reasoned that millions of people tune in to see him and most of them are Internet users, so it was a good fit. The producer showed serious interest, so our marketers, McCann Erickson, drew up a picture of Stephen's female leatherback—Stephanie Colburtle (they are females after all) holding an American flag—and I sent a formal letter to Colbert informing him that students from Drexel University who were working in Costa Rica decided to name a turtle in the Great Turtle Race after him. How could he say no now? This great honor, I told him, was in recognition of his support for saving endangered species, especially eagles. That did the trick!

The website went live April 8 with a teaser page, and 52,875 people visited it. As the race date drew closer Johnny took Leopold down to the studio to meet the cast and crew of the *Colbert Report,* and on April 11 Stephen Colbert introduced his turtle and the Great Turtle Race to the world. Colbert did a 4-minute monologue about the race, his namesake Stephanie, and several politicians that he humorously linked to turtles. The spot got rave reviews. The website immediately had a big jump in visitation, from 20,892 on April 10 to 102,794 on April 12. We had seen "the Colbert Bump." After that initial introduction, Colbert updated his audience on Stephanie's progress three more times during the race. Stephanie did very well but finally dropped off and finished second.

The Race

On April 16, 2007, my wife, Laurie, and I were standing out in the cold rain on Rockefeller Plaza in New York City with Rod Mast and his wife, Angela, their boys, and Johnny and Brooke Glidden holding signs in front of the cameras of the *Today Show.* We had gotten there at 4:30 a.m. to be up front. Rod had planned to wear his Mr. Leatherback costume (a big rubberized turtle with expressive flippers and a big smile; see www.mrleatherback.com), but the media arrangement required for that broke down and he had to stand there as a mere man. In any case, we got our five or ten seconds of fame, and there were a whopping 284,050 Internet page views of the Great Turtle Race on that day.

The website traffic was really good, considering the bad news that day. A student had shot and killed thirty-two faculty and students and wounded eighteen more at Virginia Tech. Laurie and I heard about those tragic events when we were headed for the train to return to Philadelphia. Frank Paladino called to tell us, because our son Jim was a faculty member in the Geology Department in the building next door to the shootings. We were able to call Jim in the late morning and verify that he was fine, but as acting chairman he spent his day making sure everyone in his department was safe. Those events absorbed our attention as well as that of most people in the United States. Perhaps our conservation stunt was a way people took their minds off the tragedy, but it still amazes me that the Great Turtle Race wasn't simply ignored.

Approximately 670,000 unique visitors visited www.greatturttlerace.com during the 14 days of the race. Outreach within Costa Rica resulted in national coverage before and during the race, especially in broadcasts on Channel 7 and Channel 15, along with feature articles in *La Nación* and the *Tico Times* newspapers. That coverage bolstered public and governmental support for the turtles and Las Baulas Park and helped push land acquisition actions later that spring. Some 42,871 people did more than observe—they chose a turtle, subscribing to receive daily e-mail updates during the race and to learn more about conservation issues after the race. An online petition in which individuals took a personal pledge to reduce the number of plastic bags they use garnered almost 18,000 signatures. The race raised $247,000, including $225,000 from sponsors, $18,000 from individual donors, and $4,500 from the sale of merchandise. When expenses of about $120,000 were accounted for, there was approximately $125,000 left to pay ranger salaries and other conservation costs at Las Baulas Park. Of course, there were numerous other, nonmonetary benefits.

We targeted the race at young people aged five to eighteen and their parents in the United States, as well as the broader U.S.–Costa Rican publics. In addition, we focused on reaching classrooms by adding a special Sea Turtle School page, developing a Great Turtle Race curriculum, and sending information directly to teachers. Many teachers contacted us afterward to share stories about how their classes followed the race, and to comment on what they and their students learned about sea turtles and ocean conservation.

This multimedia campaign more than doubled CI's online community. Reaching millions of people across the globe, it played a critical role in advancing the consolidation of Las

Baulas Park in Costa Rica. The success was primarily due to the imagination and hard work of Mark Breier and George Shillinger and the many people who volunteered their time and energy to make it happen. From a brilliant idea, through the early growing pains, to the final rush to the finish line, fate smiled on us all—and on the turtles. We found our celebrity spokesman in Stephen Colbert, we engaged the public and the media, and we broke new ground in conservation education. The success of the Great Turtle Race led to a Great Turtle Race II in 2008 (leatherbacks "racing" from California and Indonesia to the center of the Pacific Ocean). The Great Turtle Race II allowed us to send $75,000 to our colleagues in Indonesia to support construction of a large boat that would be used to transport them safely to the nesting beaches of Irian Jaya. In 2009 the Great Turtle Race III expanded its reach to the National Geographic Society web page and followed leatherbacks from Nova Scotia to the Caribbean in an effort supported by musicians from Pearl Jam, R.E.M., Ratatat, and others.

As far as I know, only one visitor to the site complained about the race being time-delayed. I called him personally and explained that we described the methods clearly on the website and were not trying to fool anyone. It was just a matter of compressing 6 months or more of data into a 14-day education and entertainment experience. It was like the time-delayed Olympic coverage from Japan or, as Mark had said, the West Coast delay, only longer. He seemed satisfied that there really was no other way to run the race and gave us a donation!

Two people and one idea that would not die. That's how I often describe the Great Turtle Races. Too many people credit me with the success of these endeavors, but all the credit should go to George and Mark. Do you have a great idea for sea turtle conservation? If so, don't let it die.

14

SEA TURTLES AND SATELLITES: TALES OF TECHNOLOGY

The first rays of the sun are reaching over the mountains of Costa Rica and lighting up the blue-black sea off Playa Grande. Leatherback mother #1,731 rises from the depths, moving from darkness to light as she comes to the surface. At 1 m (3 feet), she blows bubbles as she exhales in anticipation of taking a breath of clean air. She breaks the surface and with a whoosh inhales a lungful. Down for a stroke with her front flippers and back up again, she gulps a second and then third breath. The coastline is visible along with the waves and a scattering of clouds. We can see the trees along the beach and the hills beyond. She descends and heads to the bottom to await the night, when she will return to the beach to lay her clutch of sixty-five eggs. I know all this because I was with her, in a manner of speaking. This is one of the marvels brought to us from technology: I saw what she saw through a camera mounted on her carapace.

Greg's Idea

In the 1980s Greg Marshall came up with an idea: put cameras on animals and see the world as they do. Today you can scarcely flip through the channels without seeing Greg or one of his animals on TV. In the fall of 1989 I received a telephone call from

Chris Luginbuhl, the person who first got me involved with leatherback turtles off Rhode Island back in 1980. Chris said that his friend Greg had invented an underwater camera, with which he could record the behavior of leatherback turtles. This would be the first use of his invention. With some trepidation, Frank and I agreed. The last time I had worked with Chris it led to a decade of research, so we thought, maybe he's charmed.

A few weeks later Greg met us in Costa Rica with a video camera, a piece of 8-inch steel pipe, and two pieces of Plexiglas to cover the ends. That was, it seemed, the first model of the Crittercam. Greg told us he was going to make a few modifications and would then be ready to deploy the camera. Frank and I exchanged glances. There were quite a few details to work out. How will we attach it to the turtle? How will we seal the camera into the pipe? Greg said he was working on that. How will we find it when it comes off the turtle? There was no float or radio transmitter to locate the camera. How will we recover the camera at sea? We didn't even have a boat. Needless to say, we did not launch the camera on that expedition. You have to have Plan A figured out before you go, have all of your supplies and equipment ready, and then have a Plan B for backup. Greg had some more work to do, but he also learned some valuable lessons.

A decade later Chris called again. "Remember Greg Marshall?" Sure, I remembered. Nice guy with a good idea that wasn't ready for prime time. Well, explained Chris, Greg was now a producer for National Geographic and was having great success with his Crittercam on a number of animals. I had not been watching much TV at the time or I would have noticed Greg coming into public view.

Chris, who was as determined as Greg, asked if we could give it another try. By now Greg had matured in his thinking and approach. We had preliminary meetings at National Geographic, visited Greg's lab, and met the National Geographic engineers. I asked my postdoctoral fellow, Richard Reina, to take the lead in working with Greg. Greg and his assistant Kyler Abernathy developed a wonderful suction cup attachment method that he thought would hold the Crittercam to the leatherback's carapace and then release after a reasonable amount of time. It seemed the group was ready for a field test. They now had the suction cup for attachment, a camera in a waterproof housing, a float, and a radio transmitter that would reveal the location of the transmitter once it came off the turtle. So that November they deployed and recovered the camera several times on several turtles.

At the meeting of the International Sea Turtle Symposium in Philadelphia in February 2001, Greg and Richard showed scientists for the first time the underwater world of a leatherback turtle. Swimming, resting on the bottom, diving, and surfacing—the audience saw it all. They even had video of two leatherback males trying to mate with her. The film was the hit of the meeting—particularly the segment showing a male turtle smashing his head on the ocean bottom when a female avoided his amorous rush. Later, at the 2008 sea turtle meeting, Tomoko Narazaki and colleagues working with Kyler and Greg showed Crittercam footage of a loggerhead feeding on jellyfish. They were able to determine sight distances at which the turtle could see its prey.

It would have been easy for Greg Marshall to have given up. He could have chalked up his failures to youthful endeavors. Luckily, Greg kept pushing his crazy idea, and now we can see the world as a turtle sees it.

Why Molecules Matter

Wildlife genetics is a rapidly growing field that has been used quite extensively with regard to sea turtle species. In 1989 Brian Bowen, Anne Meylan, and John Avise made an important breakthrough when they used genetic studies to address an old mystery. Halfway between Brazil and Africa lies Ascension Island. It was known from tagging studies that the green sea turtles nesting on Ascension Island fed off the coast of Brazil. Why did they travel so far away when other green turtles feeding in Brazil nested in South America? It was hypothesized that the turtles undertook this 2,000-km (1,243-mile) voyage because they were returning to the place where they had hatched. And further, the hypothesis stated that they have been doing this for 65–70 million years, beginning at a time when Ascension was near the coast, and continuing to travel there as the island slowly drifted away from the coast. Fascinating, but true?

To answer the question, the Bowen-Meylan-Avise trio looked at the mitochondrial DNA of the turtles. Mitochondria occur in the cells of animals, plants, and many other organisms. They are called the cell's "powerhouse," because they fuel most of the cell's metabolism. Without getting too technical, they are important for research because they have an interesting feature—they have their own DNA.

You have trillions of mitochondria in your cells right now, fueling every heartbeat and every breath. But, the evolutionary argument goes, the mitochondria are immigrants. They live inside you with their own DNA and seem to have been, in the deep past, free-living bacteria that somehow became part of the organism they invaded. To make this tale more interesting, males do not pass on mitochondria to their offspring. The mitochondria are passed to the next generation by the mother, as they are contained in the ovum (or egg).

Brian Bowen and colleagues examined the mitochondrial DNA from the turtles nesting at Ascension Island and came to two remarkable conclusions. First, there was enough similarity among the turtles nesting there to suggest the turtles regularly returned to the beaches from whence they hatched. Second, there was not enough similarity to conclude that the green turtles had been nesting there for the past 65–70 million years. So they must have found the island more recently than that.

Bowen and many others who followed him have addressed a number of questions related to sea turtles. We now accept the idea of natal homing—that turtles generally return to the beach from which they hatched—as a rule with exceptions. The long-standing question regarding whether or not Kemp's ridleys and olive ridleys were different species seems to have been answered. They are two distinct species. We have learned that loggerheads in Mexico are distinct from those in the United States and that the loggerheads nesting in Georgia and the Carolinas are distinct from those in Florida.

Genetic studies demonstrated that 25 percent of the loggerheads that washed up dead on the shores from Virginia to Massachusetts came from Georgia and the Carolinas, 59 percent from southern Florida, and 16 percent from Quintana Roo, Mexico. We learned through Peter Dutton's genetic work that leatherbacks captured in the northern Pacific longline fisheries were from western Pacific nesting beaches in Papua New Guinea and nearby Irian Jaya, Indonesia. While completing her MS degree at the University of Florida, Sandra Encalada combined an analysis of mitochondrial DNA of green turtles with an analysis of the geological and climatic history and discovered that green turtles in the Atlantic and Pacific oceans were probably isolated from each other about 2 million years ago.

Karen Bjorndal and her colleagues examined mitochondrial DNA of green turtle colonies in the Caribbean in detail and

discovered that there was a separation in the genetics of western and eastern Caribbean nesting populations, indicating that they were distinct. They also found that, counter to what most scientists thought, small nesting populations actually contained more genetic diversity than large populations. This means that it is genetically important to protect not only a few large colonies of turtles but the many small ones as well. These data, combined with the information on natal homing, also help to explain why green turtles have not reestablished the nesting colonies that were destroyed on Bermuda, Grand Cayman, and other locations.

For many decades sea turtle biologists discussed whether there were six, seven, or eight species of sea turtles. When the "ridley turtles" were found to be two species, the question of seven or eight continued. Green turtles, some argued, were actually two species, an Atlantic and a Pacific form. Looking at bones and other features, people argued it both ways. Then the molecular data came to the rescue. Stephen Karl and Brian Bowen analyzed the mitochondrial and nuclear DNA of green turtles from around the world and found that there was only one species of green turtle in the world, although there were distinct populations of that single species.

Brian Bowen and those who came after him opened up the magical world of molecular genetics. Sometimes their explanations can be, I admit, a bit technical and somewhat boring, but Brian isn't. A devoted "Deadhead," Brian has a passion for sea turtles and music. He makes genetic data interesting by giving talks that are always filled with exciting natural history information. He is one of those passionate scientists who knows his stuff and knows how to have a good time.

Transmitters through Time

Imagine the scene: Archie Carr and the U.S. Navy attaching weather balloons to sea turtles as they tried to figure out how to deploy radio transmitters. They tried various transmitter designs, and all of them were failing. The balloon idea was supposed to help them keep track of the turtle, so they could recover the turtle and transmitter. Unfortunately, even a slight wind resulted in the balloon blowing down onto the water. They improved the balloon design by using a 4-foot-long, plastic-coated, bright-yellow, blimp-shaped balloon. They tied the radio transmitter to the balloon, moored the balloon to a float tied to a green turtle, and off they went tracking the turtle from the University of Florida research boat. They were able to receive the radio signal from as far away as 13–15 km (8–9 miles). It was a start.

Years later Howard Baldwin, from the University of Arizona, and Archie Carr did a series of experiments funded by NASA at Tortuguero and Ascension Island from 1967 to 1970. The transmitters were touchy, and in one shipment three of the four transmitters were damaged and the remaining unit only worked for 2 hours. In another attempt they attached a large torpedo-shaped metal float with a transmitter to green turtles that could be followed from a 50-foot boat for up to several hours at a time. However, the float was 1 m (3 feet) long, 15 cm (7 inches) in diameter, and weighed 9.3 kg (20 lbs). It also had a big keel to keep it upright. Needless to say, it was cumbersome and created considerable drag. It did not allow normal behavior of the turtle and sometimes got hung up on a floating log and other debris in the ocean and then came off (it was designed to release if it got snagged, to protect the turtle).

Archie was frustrated by both the state of the transmitters and the difficulty of trying to stay within 15 miles of a turtle out

at sea (radio transmitters only work at the surface of seawater, so you can only hear it when the turtle is at the surface). Sometimes hours would pass and he would wonder, "Is the turtle underwater, or did the transmitter fail again?"

In 1978, I received a call from a colleague at a campus across town telling me that Henry Prange of Indiana University was giving a seminar on green turtle physiology that afternoon. Short notice, but I jumped in the car and headed over. That was a fateful decision. The seminar was excellent, and I was inspired. At a reception afterward Henry and I talked about measuring sea turtle physiology at sea. I told him that I was familiar with telemetry and had been using it to study alligator physiology. I was sure it could work on sea turtles to measure their body temperatures in the ocean. Henry said he would try to include me in the expedition. I said that I needed to bring graduate students along to help, which he said was fine.

I called up Ed Standora, then a graduate student at the University of Georgia's Savannah River Ecology Laboratory in South Carolina. Ed had been a ham-radio-operator geek as a kid and was a wiz at telemetry. We had worked together on the alligator project and hit it off. I asked him if he would like to join an expedition to study green turtles in Costa Rica. His reply was, "Sure. What do I have to do?"

"Make us six of those multichannel sonic transmitters that you designed, and some radio transmitters, so that we can follow the turtles." I'm sure I made it sound simple and that Ed was on the other end of the phone thinking, "But those are tricky to make." But all he said was, "OK, just let me know if we are approved." I prepared a proposal with Ed's specs and asked the National Science Foundation for funds so we could join Henry's expedition. It was approved.

Sonic transmitters work by putting out a pulse that sounds like a click every time the transmitter is energized. They are particularly useful in salt water because the signal can be heard, unlike radio transmitters, when the animal is below the surface. Their range, however, is limited, typically less than a mile.

Ed designed and constructed an electronic circuit board, composed of integrated circuits, resistors, and capacitors, and he connected thermistors to it. Thermistors are variable resistors; that is, their resistance value changes with temperature. As they heat up, their resistance drops and the transmitter beeps at a faster rate. The plan was to attach thermistors to measure temperature at various points along and within the turtle's body—the carapace, the plastron, and inside the body cavity itself. Ed also designed and built small radio transmitters that we placed on small wooden floats that the turtles towed from a line attached to the back of their carapace. These would help us find the turtle from farther away, but of course, only when it was at the surface.

We arrived in Costa Rica and got to work. With Archie Carr's help we obtained an adult green turtle and set about cautiously implanting the various wires that would measure temperature. Ed and I were joined by Bob Foley, another student. We needed a boat and secured a rubber inflatable Zodiac. The turtle was released and off Ed and Bob went, crashing through the waves of Tortuguero while Archie and I cheered them on from the beach. The radio transmitters worked well, sending a signal about 16 km (10 miles) away. The information from the sonic transmitter could be read from 1.6 km (1 mile) or less away.

Ed and Bob disappeared from sight, and we waited anxiously for their return. Night came and still no sight of them. Relieved when they finally returned, we rushed to hear the tale. They had loads of body temperature data from a free-swimming green turtle! This was a first for science and was sure to attract a lot of interest. The extent of the full story would have to wait until we reviewed the data.

Archie's long years of frustration with technology had come to an end. Only human fatigue ended the tracking that day. Also, no one had been able to get any electronics to work on the black sand beach there. The sand is magnetic, composed of about 25 percent magnetite (a form of iron), and it sticks to any instruments you set down near it. To this day most sea turtle biologists refer to that short study as the first successful data collection with radio and sonic telemetry on a sea turtle. What had we learned? That green turtles heat their body by swimming and that the body temperature is different in different areas. It was a small step in unlocking the mysteries of how sea turtles live, but it inspired a lot of people with regard to the feasibility of using radio and sonic telemetry.

More and more people started to use telemetry after that day to study sea turtles. Mary Mendonca and Peter Pritchard tracked Kemp's ridley turtles in the Gulf of Mexico, and Scott Eckert used data loggers and radio transmitters to measure the movement and diving behavior of leatherback turtles at St. Croix in the U.S. Virgin Islands. Gone are the days when a ham-radio geek could build his own transmitters and wow the world. Today commercial companies provide the transmitters and the equipment to track them.

Satellite Telemetry

At 9 p.m. on June 1, 1979, D. L. Stoneburner, a scientist with the Institute of Ecology at the University of Georgia, made a technological breakthrough when he launched the first successful satellite transmitter on a sea turtle nesting on Cumberland Island, Georgia. It was an historic occasion and probably made Archie Carr smile when he heard about it. A brief time after Ed, Bob, and I had shown that radio and sonic transmitters could work, we were at the dawn of the satellite age.

Stoneburner received one transmission from the beach and another at 2:50 the next morning, when the loggerhead turtle was 4.8 km (2.9 miles) away at the northern end of Cumberland Island. The next day at 1:41 p.m. the turtle was 9.0 km (5.4 miles) away entering St. Andrew Sound, and at 2:47 a.m. on June 3 it was 16.4 km (9.8 miles) away approaching the extensive oyster bars that line the margins of the tidal marsh in the Sound.

However, that was it. Someone in a boat removed the transmitter and destroyed it that day. Still, a new technology was born. Stoneburner put out another transmitter at 2 a.m. on June 2 and two more on June 7 and July 11. Those transmitters lasted from 14 hours to 14 days and gave the first information on the movements of female loggerhead turtles during their inter-nesting period.

Stoneburner achieved this breakthrough with specially built transmitters that broadcast radio signals at the correct wavelength and were strong enough to reach a satellite in orbit overhead. The transmitters were placed in large doughnut-shaped floats that were 38 cm (15 inches) thick and 76 cm (30 inches) in diameter. He attached the closed foam-polyurethane floats to 33 m (107 feet) of rope-like wire and secured the wire to a turtle harness made of soft nylon strapping. Signals went up to the NASA Nimbus 6 weather satellite, and personnel at the Goddard Space Center in Greenbelt, Maryland, downloaded the data and calculated the locations of the turtles.

In 1980 he changed the float into a cone 8 cm (5 inches) in diameter at the end, 30 cm (18 inches) in diameter at the center, and 70 cm (42 inches) long. He tracked four more turtles for 1–34 days, obtaining important information about their movements during and after their inter-nesting period. That research reached the scientific world when he published an article in 1982, the same year that we published the results of our successful radio and sonic telemetry experiments with green turtles. We

marveled then, but now the equipment looks like something that should be on display at the Smithsonian.

In September 2009 one could see how far things had come. Scientists collaborated to reveal the tracking data on a website, www.seaturtle.org, and the result was that a person could follow in real time 159 sea turtles in 18 countries.

People often ask if the transmitters ever come off. They do. The main type of pop-up transmitter is one that records data on movements and behavior of a sea turtle and then releases from the turtle when a magnesium link dissolves. We are learning a tremendous amount about sea turtles and how to save them—we know their migration corridors, their temperature requirements, where they tend to congregate, how fast they swim, how far they dive, how long they spend at the nesting beaches, how far north and south they go, and what areas should be off-limits to fishing because there are many turtles in the area.

Jeffrey Polovina and his colleagues at the National Marine Fisheries Service (NMFS) and University of Hawaii in Honolulu mounted satellite transmitters on juvenile loggerhead turtles caught by the swordfish longline fishery in the North Pacific Ocean in 1997 and 1998. They tracked them over distances of 1,311–3,492 km (814.6–2,170 miles) as they swam westward toward Japan. These moderate-sized juveniles (41- to 81-cm [16.1- to 31.9-inch] carapace length) apparently started out near Baja Mexico, where they spent their early years after wafting over from the nesting beaches in Japan. The turtles they caught, it turns out, were on their way home. They swam against the current along two oceanic fronts. One was the Subarctic Frontal Zone, with a surface temperature of 17°C (63°F), and the other was the Subtropical Frontal Zone at 20°C (68°F). Those areas were exactly where the longline fishery set its hooks, so it was not surprising that the longliners caught large numbers of turtles. This new information suggested that seasonal or area closures of the fishery would reduce the incidental capture of loggerheads. However, officials at the NMFS did not take any action in response to Jeff's findings.

When the NMFS failed to act, the Sea Turtle Restoration Project and the Ocean Conservancy filed suit in U.S. federal court, and in 2000 Judge David Ezra ordered a large area of the North Pacific closed to longline fishing by the U.S. fleet in Hawaii. The NMFS fought to reopen the fishery. Every time the agency proposed to resume longline fishing the conservation groups sued again. Lawyers and judges kept busy for several years. It appeared that the NMFS personnel in Hawaii operated in the same way that their colleagues did in the east with regard to shrimp trawling and turtle excluder devices: fishing first and sea turtles last.

Some of the most interesting findings for conservation from the use of satellite telemetry have come from the application of satellite transmitters to leatherbacks in the Pacific Ocean. Frank Paladino, Stephen Morreale, Ed Standora, and I put eight satellite transmitters on leatherbacks from Las Baulas Park in 1990 to 1994. Over 4 years the leatherbacks all headed southwest toward the Galápagos Islands when they finished nesting. They stopped by Cocos Island and reached the Galápagos before heading into the South Pacific. Scott Eckert launched several satellite transmitters on leatherbacks nesting in Mexico. Most also headed southwest and passed the Galápagos, and a few stayed along the equatorial upwelling west of there. Our data indicated that there was a migration corridor from Costa Rica to Cocos Island and the Galápagos. That led Conservation International (CI) to launch the Eastern Tropical Pacific Seascape (ETPS) project in 2004, in which Costa Rica, Panama, Colombia, and Ecuador agreed to multinational coordination of marine resources in their exclusive economic zones. This project provides hope that sea turtles, sharks, and other animals in this huge area of

ocean—from Costa Rica to 200 miles west of the Galápagos, to the coast of Ecuador, and back to Costa Rica—will finally receive protection from overfishing, shark finning, gill-netting, and other activities that, unchecked, will eventually destroy the region's populations of sea turtles.

15

2100: A WORLD WITH, OR WITHOUT, SEA TURTLES?

The modern sea turtle conservation movement began in the 1950s, when Archie Carr summarized all that was known about these animals in his *Handbook of Turtles.* His later books, *The Windward Road: Adventures of a Naturalist on Remote Caribbean Shores* and *So Excellent a Fishe: A Natural History of Sea Turtles,* enchanted readers and called the world to action to save the dwindling populations. Archie founded the Caribbean Conservation Corporation (CCC), which has served as an inspiration and a model for many other conservation organizations dedicated to saving sea turtles. It gives one hope to rattle off the official-sounding names like: the Sea Turtle Restoration Project, PRETOMA, ARCHELON, Projeto TAMAR (TAMAR is an abbreviation of *tartarugas marinhas,* or "sea turtles"), Wider Caribbean Sea Turtle Network (WIDECAST), and MEDASSET. There are many others, some more effective than others, but all with the same goal: preventing and reversing the decline of sea turtle populations. The world's larger conservation organizations, such as Conservation International, the Wildlife Conservation Society, the Nature Conservancy, and World Wildlife Fund, are all extremely active in sea turtle conservation. Each year the International Sea Turtle Society brings together scientists and

conservationists to discuss the latest information on sea turtles and to develop plans to restore their numbers.

Yet with all the dramatic and often heroic efforts, sea turtles still hang close to the edge of extinction. Without these efforts there is little doubt that things would be much worse. Kemp's ridleys would certainly have gone extinct. There would not be any green turtles in the Caribbean. Most other sea turtle populations would be gone, or at least reduced. There would certainly not have been the recoveries we have witnessed in some areas, such as population growth at Tortuguero in Costa Rica and the increases on Ascension Island. There are reasons for celebration. The year 2000 came and went, and the world had the same number of sea turtle species as it had in 1900. There were, to be sure, many fewer turtles in the interceding 100 years (we had lost perhaps 90% in terms of overall numbers in that 100-year span). Sea turtles are a shadow of their former populations in most of the world. That is a sad fact of life. Still, we have the seven species, still alive in the wild and still good candidates for recovery.

With Tortuguero held out by so many as a success story, some might find it shocking that in 2009 the administration of Costa Rican president Óscar Arias proposed carving up Tortuguero National Park to favor development. The effort mirrored those that sought to undo Las Baulas Park. Until 1982, Ascension Island turtles seemed remote and protected, but then the British military showed up and started taking sand from the beaches during the Falklands War. In 2010, the Great BP Oil Spill wiped out thousands of sea turtles and much of their food in the Gulf of Mexico, undoing a decade of urgent conservation work.

Sea turtle conservationists, at least those who have been at it for a few years, know the next threat is either coming up the walk or right around the corner. The olive ridley *arribada* in

India—one of the world's most amazing events—is threatened because of fishing and port development. Longline and gill-net fishing are harming loggerheads and leatherbacks in the Pacific. Loggerheads in Florida are declining despite the Archie Carr National Wildlife Refuge and the effective implementation of turtle excluder devices.

Have we accomplished much or little? The answer is yes to both. Those now engaged in the struggle have much to be proud of, and there is plenty of work for the next wave of sea turtle biologists and conservationists.

Mario Boza and Kenton Miller were once the next wave. Mario got excited by a university-level course in wildlands management in which he went to the Great Smoky Mountains National Park in the United States. Something clicked for Mario, and he made it his mission to create a national park system in Costa Rica. There were only a couple of protected areas in the country at the time and one of them was only a park on paper, not in any real sense of the word. Mario was joined by his friend Álvaro Ugalde, who had benefited from a U.S. national parks training course funded by Archie Carr through the CCC. Together Mario and Álvaro worked with Doña Karen Olsen, the wife of then-president José (Don Pepe) Figueres. Through these efforts parks were established at Tortuguero, Poas Volcano, and Cahuita in 1970. The next year they formed Santa Rosa National Park. They used every means they could to accomplish their goal of a national parks system, including fundraising from international sources, domestic funding, clever use of U.S. Peace Corps volunteers, and creating dialogue with the media and citizens. In 1972 the Costa Rican Parks Department had a staff of twelve Ticos and eighteen Peace Corps volunteers. Almost four decades later, Mario and Álvaro are still working hard. Neither has gotten wealthy from their efforts. Both live in modest houses, and Mario still takes the bus to work. Álvaro is

still working with local communities in watershed protection in the north-central area of the country. Mario is working on watershed protection and water pollution problems related to tourist development in the Central Pacific zone. Both are still fighting for consolidation of the parks, and so is first lady Doña Karen.

Mario and Álvaro jump-started the economic engine that runs Costa Rica today—parks and ecotourism—and remain the conscience of Costa Rica. Along the way many international friends and organizations have helped them, including Archie Carr. Many students and volunteers have walked the beaches to study and protect what Mario and Álvaro have built. Many have faced poachers armed with machetes. Many have helped hatchlings to the ocean.

Sea turtles still exist today because there is a large family of people who take time out of their lives, or even dedicate a good portion of their lives, to study and save those animals. We have not saved everything, but we have saved quite a lot, considering the odds that we have faced. With your help we can save even more.

The Danger of Compromise

There is an old saying in American politics: "An honest politician stays bought." In other words, when a politician makes a deal, he should stick to it. Unfortunately, land developers follow different rules. Whenever someone proposes to make a sea turtle beach a protected area or park, there are always other people who quickly object because they see it as a good place to put a hotel or houses. Then a discussion ensues in which the two sides try to find a reasonable solution. The developers complain about all the money they imagined that they would have made and for which they now need to be compensated. Eventually, the politicians prevail upon the conservationists to be reasonable and take three-quarters or half of a loaf. That is all the politicians can get and that will have to do. With luck, at least half of the area that is needed will become protected. And compromise is, after all, the essence of a democratic society.

That is never the end of the story. After a few years, as development proceeds outside the park, pressure builds for a new compromise. "The park is too big" comes first, then the argument that there is too much economic hardship for the people, and if land purchases are needed over time (say a decade or more) the money to buy the land always seems to disappear. Another compromise is proposed and perhaps accepted. Then in a few years another compromise is needed. So it would go on until the park is whittled away to nothing. The developers, who agreed to the terms when the park was established, do not stay bought. They always want to take another piece of the pie, because there is money to be made. Making money is a powerful motivation, but the developers know something else as well. First, the politicians often take their side. Second, the conservationists often cave in. The latter is something we can stop. The price of conservation, like liberty, is eternal vigilance.

Las Baulas Park in Costa Rica provides an excellent example of this lesson. In 1990, when Mario Boza, Frank Paladino, María Teresa Koberg, and I were discussing a law to save the beaches for the leatherback turtles, we had as a model a proposal by Peter Pritchard called—can you believe it?—"For All Time." We thought that our proposal had to be "reasonable." We knew that Tamarindo, then a small village, would not be included in the park, so it was never put in the proposed law. When the proposal was passed in 1995 and signed into law, it did not include Tamarindo and its beach, as expected. We lost quite a bit of the land we had pushed for, but we had the park and it was in law. We had made big compromises, and saved, we reasoned,

what we could. Since that time the developers have tried time and again to shrink the park. First, they debated the wording of the law. More recently, they convinced President Óscar Arias to offer a bill to the Costa Rican Congress eliminating much of the land in the park.

I cannot tell you how many times I have been told "you must compromise" and accept a portion of the new proposal. They say that we should "work with" the local community and that the people who want to live on those lands are really nice and want to save the turtles, too. They say local people need jobs. They urge us to accept the compromise with the developers. My colleagues and I have learned that we have to present a unified front and try to point out that the size of the original park was already a big compromise with development.

Once a protected area has been defined, you have to protect it in its entirety or it will suffer the death of a thousand cuts. What is now protected at Las Baulas is the minimum needed to save the turtles. Any additional compromise would be the beginning of the end for them. What can be done at Las Baulas Park can be done at Tortuguero and around the world. Literally, hold your ground.

It Is Up to You

There are three generations of sea turtle biologists and conservationists alive today. The oldest generation still remembers working with Archie Carr. The middle generation has heard the stories about Archie. The new generation is just getting started on nesting beaches around the world. It is the latter two generations who will or will not see to it that the seven species of sea turtles make it to the year 2100.

Until recently scientists could still hope to find new colonies of sea turtles and new nesting beaches. Peter Pritchard, for example, found Playa Grande in Costa Rica in the 1980s, and Scott Benson and Peter Dutton discovered leatherback beaches in Irian Jaya, Indonesia, in the 1990s. Today I do not think there are any new nesting beaches to discover. What we know about is probably all we have in the world, and we may lose some of these despite our best efforts, simply because of sea-level rise resulting from global warming.

There are going to be a lot more people coming into the world over the next few decades. Almost certainly the world's population of humans will reach 10 billion. They will want houses and vacation homes. They will love the beach as you do. The pressures on nesting beaches and sea turtles in the oceans will increase. Does the next wave of sea turtle biologists and conservationists have the will to engage in the long and difficult struggle? I know this much: the developers have the will to develop, the commercial fisheries have the will to take all they can, and the poachers have the will to harvest all the eggs they can carry.

Whatever the scope of the effort, it is ultimately local conservation, working with local communities, that is the best way to save sea turtles. You cannot save them in an area unless you engage the people who live with them. Building strong relationships with the people who share the land and sea with the turtles changes the way the government looks at the situation. Plus, it is the right way to address the problem. That is, social and economic justice are important parts of an effective conservation plan. It is neither fair nor possible to expect local people to bear the entire cost of saving a sea turtle population. If they rely on turtle eggs or fishing methods that catch turtles, you have to help them find a substitute source of income so that they can live a decent life, raise their families, and have hope for a better future. For example, indigenous people in the Kei Islands in the Pacific Ocean off Irian Jaya used to exploit

leatherback turtles in a ritualistic manner and considered them sacred. But both an increase in human population size and a breakdown in tradition led them to adopt a commercial approach to leatherback exploitation. A new paradigm is needed to protect turtles there, perhaps encouraging a return to the less destructive sacrificial ritual.

Sea turtles need a constituency where they live. Arguably the most effective long-term efforts have been to teach children to love sea turtles. When they grow up, they will become advocates for the turtles.

What is your part in this unfinished play? You can walk the beach. You can volunteer locally or internationally. You can get lights turned out on nesting beaches. You can keep that hotel from going up on the turtle beach. You can join local, national, or international organizations that work to save turtles. Ed Drane helped save sea turtles in South Carolina, Georgita Ruiz saved olive ridleys and green turtles in Mexico, and Randall Arauz saved them in Costa Rica. They are not superhuman. In fact, they are ordinary people who just did what they could to make the world a better place. Dream of an ocean where turtles will swim free of longlines and gill nets and come ashore to nest on dark beaches without poachers. I believe that, despite all of the problems and all of the threats, we will save the world's sea turtles. We will do it one beach at a time, one turtle at a time, and one person at a time. You need to be that person. Come join us.

INDEX

Abernathy, Kyler, 186
Ackerman, Ralph, 20
ADIO (Asociación de Desarrollo Integral de Ostional), 153, 155, 157, 158; tasks of, 151–52, 154
animals, experimental. *See* experimental animals
Arauz, Randall, 11, 68, 85, 112, 135, 205
ARCHELON (Sea Turtle Protection Society of Greece), 11–12, 114, 198
Archie Carr National Wildlife Refuge, 107, 165, 200
Arias, Óscar, 85, 141, 142–43, 199, 203
Ascension Island, 187, 188, 199
Atlantic Ocean, 66, 67, 163, 188, 189; Gulf Stream, 62–63; longlines in, 88, 89; North Atlantic Gyre, 63–64; sea turtle population, 12
Atomic Energy Commission (AEC), 98
Australia, 12, 66, 67, 70, 146, 163; natural predators in, 48–49
Avise, John, 187
Azores Islands, 61–62, 66

Bailey, Lisa, 177
Baldwin, Howard, 190
beaches: conservation efforts around, 14, 108–10, 114–16; and development, 4, 46, 50, 106–7, 108, 111, 115; and lighting, 4, 49–50, 108, 111–12, 115, 205; and nesting, 49–50, 100, 101–2, 203–4; and noise, 50–51; water beneath, 19–22. *See also individual beaches*
Benson, Scott, 204
Bioko Island, 84
Bjorndal, Karen, 12, 61, 65, 188–89
Blackbeard Island, 108
Blanco, Gabriela, 5–6, 7, 8, 9
Block, Barbara, 175, 176
Bolten, Alan, 61, 62, 65
Boone, "Sinkey," 79, 80–81
Bowen, Brian, 187, 188, 189
Boza, Mario, 133, 137, 202; creates national parks and ecotourism, 200, 201; as vice minister of environment, 134, 136
BP oil spill, 2–3, 199
Bravo, Hernan, 135–36
Breir, Mark: and Great Turtle Race, 172, 173, 174, 175, 178–79, 183; marketing experience of, 169, 170, 171
Brongersma, Leo, 58–59
Burke, Vincent, 70
Bush, Jeb, 106–7

Calderón, Rafael Ángel, 134
Camhi, Merry, 28, 29, 30
Campbell, Lisa, 153, 154, 158
Cape Cod Bay, 71
carbon dioxide (CO_2), 20, 22; and global warming, 161–62, 166–67
Caribbean Conservation Corporation (CCC), 198, 200
Carr, Archie, 35, 49, 61, 117–18, 200; as conservationist, 14, 104, 198, 203; on natural predators, 45–46; opposition of, to sea turtle farms, 59; on sea turtle migration, 53, 57; and telemetry, 190, 192, 193; on TSD, 27, 28; works: *Handbook of Turtles*, 198; "Rips, FADS, and Little Loggerheads," 60; *So Excellent a Fishe*, 45–46, 57–58, 198; *The Windward Road*, 198
Castro, René, 136
Chaves, Amy, 156
Chesapeake Bay, 70–71
China, 85
CIMACO (Ostional National Wildlife Refuge Inter-institutional Consulting Council), 157
Clinton, Bill, 101, 136–37
coatis, 15, 48, 103, 104
Colbert, Stephen, 180, 181, 183
coloration, 58
conservation efforts: anti-poaching, 7–8, 119, 126, 127–28, 134; by Archie Carr, 14, 104, 198, 203; challenges of, 204–5; danger of compromises in, 201–3; by Earthwatch, 119, 135, 137; ecotourism, 128–31, 153–54, 157, 201; education, 14, 130, 205; ETPS project, 140, 170–71, 196–97; Great Turtle Race as, 169, 171, 173, 178, 182–83; and greed, 4; for green turtles, 104–5, 199; for Kemp's ridleys, 28, 199; lawsuits, 85, 90, 104, 142, 196; by Leatherback Trust, 111, 138–40; list of organizations, 198–99; for loggerheads, 11–12, 108; and NMFS conflict of interest, 90, 95, 97, 98–99, 196; population-monitoring, 107, 126, 157–58; at Tortuguero, 104–5, 108, 118, 199; toward beaches, 14, 108–10, 114–16; what can be done, 2, 14, 115–16, 166–67, 205. *See also* Las Baulas Park; wildlife refuges
Conservation International (CI), 140, 196
convergence zones, 3, 59–60, 164
Cornelius, Steve, 146
Costa Rica: authorizes Ostional egg harvesting, 151; bans egg poaching, 150; Coastal Current around, 68–69; conflicts of interest in, 84–85, 98, 110–15, 136, 141–43, 199, 202–3; ecotourism in, 128–30, 153–54, 157, 201; egg poaching in, 4–8, 10, 34–35, 123–26, 151, 159; environmental groups in, 139–40, 142, 143; and Great Turtle Race, 177, 182; Guanacaste Coast, 68, 84, 132–33, 134, 144, 159; media in, 134, 136, 143, 182; Ministry of

Environment and Energy, 134, 136, 155–57, 177; national park system, 45, 49, 110, 126, 132, 200, 201; National Sea Turtle Program of, 134; Nicaraguan immigrants in, 48; "Peace with Nature" initiative, 141, 142; SALA IV in, 142–43. *See also* Las Baulas Park; Ostional egg harvesting; Playa Grande; Tamarindo; Tortuguero
Cotroneo, Laurie, 41
Crowder, Larry, 89
Cumberland Island National Seashore, 108
Curtis, Steve, 150

Delaware Bay, 71
dinosaurs, 125
Diriá Forest National Wildlife Refuge, 144
Dodson, Peter, 125
dogs, 44–46, 123
Domingo, Andres, 88–89, 90
Dornfeld, Tera, 56
Drake, Dana, 41–42
Drane, Ed, 109–10, 205
dredging harbors, 97
Drexel University, 126, 135, 177, 179
driftlines, 59–60, 75
Dutton, Peter, 188, 204

Earthwatch, 119, 135, 137
Eastern Tropical Pacific Seascape (ETPS), 140, 170–71, 196–97
Eckert, Scott, 68, 196
ecotourism, 128–31, 153–54, 157, 201
egg poaching: in Costa Rica, 4–5, 6–8, 10, 34–35, 123–26, 151, 159; efforts to stop, 7–8, 119, 126, 127–28, 134; impact of, on sea turtle populations, 32–33; in Mexico, 127–28; organization of, 124, 125; and Ostional experiment, 151, 159
eggs: absorption of oxygen by, 18, 19–20, 40; burial of, 18–22, 25, 29; embryo development inside, 15–18; hatching of, 36–38; hydration of, 18, 19; incubation period for, 15, 22, 26; natural predators of, 15, 102–4, 106, 123; seen as sexual-performance enhancers, 34–35; sex determination of, 11, 26, 27–32; temperature of, 19, 22–26, 36–37; and tidal flooding, 20, 21, 26. *See also* nesting
Ehrenfeld, David, 28, 117–18
Eisenstaedt, Alfred, 73–74
El Niño, 146, 162–63, 164
Encalada, Sandra, 188
Endangered Species Act, 79, 90
Epperly, Sheryan, 83
experimental animals, 30–31
extinction: as danger, 1, 12, 199; measures to prevent, 91

feeding, 67, 69, 70, 71, 107, 173; by hatchlings and juveniles, 62–63, 69; ingestion of human debris, 75, 76, 107
Figueres, José María, 136, 137
Finding Nemo, 62
fishing, 78–92; circle hooks, 90, 91; gill nets, 93–95, 107, 200;

fishing (*cont.*)
international shrimping, 84–85; longlines, 10, 87–90, 92, 200; marine reserves and, 91; oceanic drift nets, 94; old methods of, 85–87; pound nets, 96–97; scallop dredging, 96; shrimp trawling, 79–84, 92, 107; and TEDs, 79, 80–83, 84, 85, 91, 107; voluntary regulation of, 79, 91
flatbacks (*Natator depressus*), 12, 67, 70
Flesh on the Bones, 126
Florida, 53–54, 66–67, 71, 83, 188; declining sea turtle population in, 107–8, 200; development in, 106–7, 165; as nesting site, 22, 31, 107–8, 160; wildlife refuges in, 50, 107, 165, 200
Foley, Bob, 192
Fonseca, Luis G., 41
Ford, Harrison, 171
Frick, Jane, 53
frigate birds, 41–42, 48, 60, 104

Galápagos Islands, 140, 170, 171, 197
Gaylord, Murray, 175
genetics, wildlife, 187–89
Georgia, 82, 83, 193; nesting beaches in, 62, 108, 160, 188; wildlife refuges in, 108–9
Glidden, Brooke, 179–80
Glidden, Johnny, 179–80
global warming, 10–11, 160–62, 166–68; effect of, on sea turtles, 11, 162–65
Great Turtle Race, 169–83; conservation awareness as goal of, 169, 171, 173, 178, 182–83; corporate sponsorship for, 171, 174, 176–77, 179, 180; idea for, 170–71; money raised from, 182, 183; and outreach within Costa Rica, 177, 182; and subsequent races, 183; time-delayed coverage of, 175, 183; website for, 177–78, 180–82
Greece, 114–15
green turtles (*Chelonia mydas*), 5, 10, 67, 104; conservation efforts around, 104–5, 199; feeding, 67, 69; genetics, 188–89; hatchlings, 43, 69; longline threat to, 10, 88–89; nesting, 19, 107–8, 123; population decline, 12, 117
Guanacaste, 68, 84, 110, 132–33, 144, 159
Gulf of Gabon, 84
Gulf of Mexico, 12, 66–67, 103; BP oil spill in, 2–3, 199
Gulf Stream, 60, 62, 64, 65, 67

Handbook of Turtles (Carr), 198
Harrington, Dave, 82
hatchlings: drinking of water by, 52; emergence from eggs, 17, 38–40; feeding, 62–63, 69; and heat, 40–44; and light, 49–50; "lost years" of, 57–59; magnetic guidance system of, 53–55; migration by, 53–54, 62–68, 69–70; mortality rate of, 56; and noise, 50–51; predators of, 44–49; rate of growth of, 69; "surfing" by, 62–65. *See also* eggs; nesting

hawksbills (*Eretmochelys imbricata*), 12, 69
Hays, Graeme, 41
Hendrickson, John, 22–24, 42
Hersey, Wanny, 172
Hilton Head Island, 109–10, 115
Hirth, Harold, 117–18
Honarvar, Shaya, 41
Hughes, D. A., 144–45
human debris, 74–77, 107
Hutchinson, Brian, 177
Hyslop, John, 145

INCOPESCA, 98, 157
India, 85, 199–200
integrated circuits, 25
Inter-American Tropical Tuna Commission (IATTC), 91
International Sea Turtle Society, 139, 198–99
International Sea Turtle Symposium (2001), 186

Jackson, Jeremy, 12
jaguars, 48, 104, 105
Japan, 65, 66, 195
Jekyll Island, 108–9, 115
Jupiter Island, 106, 107

Karl, Stephen, 189
Kei Islands, 204–5
Keinath, John, 71, 72, 73
Kemp's ridleys (*Lepidochelys kempii*), 3, 88; conservation efforts around, 28, 199; feeding, 69, 70; genetics, 188; migrations, 66–67, 70, 71; nesting, 18, 103, 122; population decline, 12, 78
Kiawah Island, 50, 108
Knorr, Joseph, 94
Koberg, María Teresa, 7–8, 128, 133–35, 202

Laganas, 114
Las Baulas Park, 49, 132–43; conservation work of, 128, 134–35, 137; creation of, 128, 132, 134–35, 136–37, 202; efforts to undercut, 132, 136, 137, 141–42, 143, 199, 203; fight by environmental groups around, 139–40, 142, 143; and Great Turtle Race, 177, 182–83; land acquisition for, 140–42; as leatherback nesting beach, 33, 132, 138; Leatherback Trust and, 138–40; media attention to, 134, 136, 143; SALA IV on, 142–43
leatherbacks (*Dermochelys coriacea*), 76, 203–4; feeding, 70, 71, 173; fishing practices as threat to, 83, 88, 89, 200; genetics, 188; hatchlings, 26, 43, 55, 67–68; at Las Baulas, 132, 138; migrations, 68–69, 71; nesting, 19, 107–8, 122–23, 126; population decline, 12, 32, 93, 173, 200; at Tamarindo, 110–11, 112–13
Leatherback Trust, 111, 138–40, 141, 173, 174
Lennox, Linda, 145, 146, 147
Leslie, Alison, 44–45, 46
Lewison, Rebecca, 89, 90
Limpus, Colin, 146
List, Al, 135
loggerheads (*Caretta caretta*), 3, 46,

loggerheads (*cont.*)
61, 66; conservation efforts around, 11–12, 108; feeding, 69, 107; genetics, 188; longlines as threat to, 88, 89; migrations, 62–64, 66, 69, 70–71; nesting, 19, 107, 108, 114; population decline, 78, 200
Lohmann, Cathy, 54–55, 63–64
Lohmann, Ken, 54–55, 63–64
Long Island Sound, 70
Louisiana, 80, 81
Luginbuhl, Chris, 72, 76, 185
Lura, Jessica, 172
Lutcavage, Molly, 118
Lux, Jenny, 76–77

Madeira, 66, 67
magnetic orientation, 54–55, 64–65
Malaysia, 32
Margaritoulis, Dimitris, 11–12
marine reserves, 84, 91
Marine Turtle Recovery Team, 80
Marshall, Greg, 184–86
Martins, Helen Rost, 61, 62
Mast, Rod, 139–40; and Great Turtle Race, 169, 171, 174, 175, 176, 181
McCann Erickson, 180
McCollum, Andy, 55–56
MEDASSET (Mediterranean Association to Save Sea Turtles), 114, 198
Mendonca, George, 72–74
Mendonca, Mary, 193
metabolic heating, 25, 26
Mexico, 65, 66, 127–28, 195
Meylan, Anne, 187
migration: by birds, 53; by hatchlings, 53–54, 62–68, 69–70; of Kemp's ridleys, 66–67, 70, 71; of leatherbacks, 68–69, 71; of loggerheads, 62–64, 66, 69, 70–71; magnetic orientation in, 54–55, 64–65; of olive ridleys, 67
Miller, Kenton, 200
Monterey Bay, Calif., 70
Moore, Ken, 172
Moore, Kris, 172
Moran, Kate, 41
Morreale, Stephen, 6, 28, 36, 70, 123, 196
mortality rates, 56, 78, 89, 91, 93
Mrosovsky, Nicholas, 32
Murillo, Grettel A., 41
Murphy, Sally, 80–81, 82, 83
Murphy, Tom, 83
Musick, Jack, 70–71

Narazaki, Tomoko, 186
National Air and Space Administration (NASA), 25, 190
National Marine Fisheries Service (NMFS), 91, 96; conflict of interests of, 90, 95, 97, 98–99, 196; and TEDs, 79, 80, 81, 82–84
natural selection, 19–20, 64
Nature Conservancy, 198
nesting: *arribada* system, 15, 18–19, 102, 103, 106, 151–52, 155; and bacterial contamination, 145; and ecotourism, 129; of green turtles, 19, 107–8, 123; of Kemp's ridleys, 18, 103, 122; of leatherbacks, 19, 107–8, 122–23,

126; of olive ridleys, 18, 103, 122, 145; process, 118–19, 120–23; rain and, 19, 22, 26, 32. *See also* beaches; egg poaching; eggs
New York, 70, 71
Nicaragua: Contra War in, 138; immigrants from, 48
Nixon, Richard, 79
North Atlantic Gyre, 63–64
North Carolina, 80
Nova Scotia, 71
Nuclear Regulatory Commission (NRC), 98

Ocean Conservancy, 80, 91, 196
oceans: convergence zones and driftlines, 3, 59–60, 75, 164; light of, 49; noise of, 50; oceanic zones, 65, 67; salinity of, 52; tidal flux, 20, 21; wind streams, 55. *See also* Atlantic Ocean; Pacific Ocean
O'Connor, Mike, 118, 161
Ogren, Larry, 53, 117–18
olive ridleys (*Lepidochelys olivacea*), 43, 125; feeding, 67, 69; fishing practices as threat to, 10, 85, 88, 199–200; genetics, 188; migrations, 67; nesting, 18, 103, 122, 145; population decline, 12
Olsen, Karen, 200, 201
Operation Green Turtle, 104
Orrego, Carlos Mario, 7, 156–57, 158
Ostional egg harvesting, 144–59; arguments for, 151, 154, 159; biological impact of, 155–56, 158–59; economic interests in, 151, 152–53, 154–55; legal authorization for, 151–52; as stain on conservation, 158–59; villagers and, 148–49, 152–53
Ostional National Wildlife Refuge, 150–51, 155–56, 157
Oyster Creek nuclear power plant, 97

Pacific Ocean, 60, 65, 66, 67; convergence zones in, 60, 164; El Niño and, 146, 162–63, 164; ETPS conservation zone, 140, 170–71, 196–97; global warming and, 165; Great Turtle Race in, 173, 175, 183; green turtles in, 9–10, 146, 188, 189; leatherbacks in, 138, 143, 164, 173; longline fishing in, 89, 91, 188, 196, 200; pollution and human debris in, 75, 201; sea turtle population, 12; winds, 55
Padilla, Clara, 133, 139
Paladino, Frank, 3, 112, 118, 161, 163, 181, 202; educates ecotourism guides, 130, 135; and egg poachers, 6, 124–26; and Las Baulas Park, 133, 137, 141, 142, 143, 147; and satellite transmitters, 185, 196
Panagopoulou, Aliki, 11–12
Papadakis, Constantine "Taki," 177
Pearson, Steven, 41
Pieau, Claude, 27
Piedra, Elizabeth, 5
Piedra, Rotney, 5–6, 88, 137, 177, 179
Pinckney Island, 108, 110

plastics, 75–77, 107, 173
Playa Grande: antipoaching efforts at, 7–8, 128, 133–35; ecotourism at, 128, 130, 157; and global warming, 162, 165; and Las Baulas Park, 128, 130–33; lights and development at, 49, 111–12; nesting and hatchling research at, 20–22, 37–38, 56, 76; as nesting site, 111, 204; predators at, 45, 48, 102–3, 105
Playa Nancite, Costa Rica, 26, 29, 42, 103–4, 145
Playa Ostional. *See* Ostional egg harvesting
Playa Ventanas, Costa Rica, 105, 111
Plotkin, Pamela, 111
pollution, 111, 116
Polovina, Jeffrey, 195
populations, 12, 199; of flatbacks, 12; of green turtles, 12, 117; impact of poaching on, 32–33; of Kemp's ridleys, 12, 78; of leatherbacks, 12, 32, 93, 173, 200; of loggerheads, 78, 200; monitoring of, 107, 126, 157–58
power plants, 97
Prange, Henry, 191
predators, 15, 44–49, 102–4, 106, 123; coatis, 48, 103, 104; dogs, 44–46, 123; frigate birds, 41–42, 48, 60, 104; jaguars, 48, 104, 105; raccoons, 46–47, 48, 102–3, 123; vultures, 18, 49, 103, 104
PRETOMA (Programa Restauración de Tortugas Marinas), 11, 99, 198
Pritchard, Peter, 80, 193, 202, 203–4; and Las Baulas Park, 133, 134; on Ostional egg harvesting, 151; *Time* magazine tribute to, 11
Projecto Tamar, 198

Quirós, José, 135

raccoons, 15, 46–47, 48, 102–3, 123
Rancho Nuevo, Mexico, 3, 28, 31–32, 66, 128
Reina, Richard, 186
reptiles, 16
Reynolds, David, 105
Rhode Island, 71
Richard, J. D., 144–45
Robinson, Doug, 147, 149, 151, 152, 154, 159
Rodríguez, Esperanza, 7, 8, 133–34
Ruiz, Georgita, 28, 29, 30, 127–28, 205

SALA IV, 142–43
Salmon, Mike, 53, 54
Santa Rosa National Park, Costa Rica, 49, 103, 200
Santidrián Tomillo, Pilar ("Bibi"), 32–33, 45, 47, 56
Sargassum, 3, 60, 61, 63, 67
satellite telemetry. *See* telemetry
Savannah Coastal Wildlife Refuge, 110
sea turtle farms, 59
Sea Turtle Restoration Project, 196, 198
Sea Turtles (Spotila), 114

Seminoff, Jeff, 8, 9
sex determination. *See* temperature-dependent sex determination
Shillinger, George, 140; and Great Turtle Race, 173, 174, 175, 176, 183; and satellite transmitters, 163–64, 169, 170–72
Shoop, Bob, 72
Sieg, Annette, 120
snapping turtles (*Chelydra serpentina*), 27
So Excellent a Fishe (Carr), 45–46, 57–58, 198
Sotherland, Paul, 20–22
South Africa, 59, 130
South Carolina, 50, 81, 83, 108
Spotila, Jim, 181–82
Standora, Ed, 70, 71–72; at Playa Ostional, 145, 146, 149, 150; on sex determination, 27, 36; and telemetry, 191, 192, 196
Stone, Lisa, 172
Stoneburner, D. L., 193–94
symbiosis, 104

Taiwan, 84–85
Tamarindo, 110–14; ecotourism at, 128–29; and Las Baulas Park, 132, 134–35, 202; lights and development at, 49, 111–12, 113
Teas, Wendy, 83
telemetry: radio and sonic, 73, 190–93; satellite, 5–6, 173, 176, 179, 184–87, 193–97
temperature-dependent sex determination (TSD), 11, 26, 27–32
Terengganu, 32, 33
thermistors, 24
thermocouples, 23–24, 25
thermoregulation, 68
TOPP (Tagging of Pacific Pelagics), 175–76
Tortuguero, 53, 75; as green turtle nesting site, 104, 117–18, 164; as national park, 49, 199, 200; natural predators at, 45–46, 48, 104, 105; poaching at, 104, 150; successful conservation efforts at, 104–5, 108, 118, 199; temperature-determination experiments at, 26, 28–29, 36
turtle deflector device (TDD), 96
Turtle Excluder Device (TED), 79, 80–83, 84, 85, 91, 107
Tybee Island, 108

Ugalde, Álvaro, 200–201
United Nations, 94, 140
U.S. Agency for International Development (USAID), 138
U.S. Army, 101, 115

Valverde, Roldán, 157, 158
Virginia Polytechnic Institute and State University, 181–82
Virgin Islands, 70
vultures, 18, 49, 103, 104

Wallace, Bryan, 139, 179
Walton, Rob, 171
Wassaw Island, 108
water: beneath beaches, 19–22; rain, 19, 22, 26, 32; sea turtle drinking of, 52. *See also* oceans
Wheatstone bridge, 24

Wider Caribbean Sea Turtle Network (WIDECAST), 198
Wildlife Conservation Society, 198
wildlife refuge(s), 50, 101, 111; Archie Carr National, 107, 165, 214; Diría Forest National, 144; in Georgia and South Carolina, 108; Ostional National, 150–51, 155–56, 157
Windward Road, The (Carr), 198
Wishrad, Vinnie, 177
Witherington, Blair, 3, 53–54
Wolf Island, 108
Woody, Jack, 27–28
World Wildlife Fund, 91, 126, 198
Wyneken, Jeannette, 53, 54

Yahoo, 171, 175, 176, 177–78
Yntema, Chester, 27
Young, Andrea, 38

Zakynthos Island, 114–15
Zandona, Eugenia, 120